CHANGING CONTOURS
OF WORK

To Jeffrey and Jonathan Sweet, two hardworking family men.
-Stephen Sweet
Ithaca, New York

To the working students of Cleveland State University, who are living the problems described in this book.
-Peter Meiksins
Cleveland, Ohio

SOCIOLOGY FOR A NEW CENTURY

CHANGING CONTOURS OF WORK

Jobs and Opportunities in the New Economy

◆

STEPHEN SWEET
Ithaca College

PETER MEIKSINS
Cleveland State University, Ohio

 PINE FORGE PRESS
An Imprint of Sage Publications, Inc.
Thousand Oaks • London • New Delhi

For information:

Pine Forge Press
A Sage Publications Company
2455 Teller Road
Thousand Oaks, CA 91320
E-mail: order@sagepub.com

Sage Publications Ltd.
1 Oliver's Yard
55 City Road
London EC1Y 1SP
United Kingdom

Sage Publications India Pvt. Ltd.
B 1/I 1 Mohan Cooperative
Industrial Area
Mathura Road, New Delhi 110 044
India

Sage Publications Asia-Pacific Pte. Ltd.
33 Pekin Street #02-01
Far East Square
Singapore 048763

Printed in the United States of America.

Library of Congress Cataloging-in-Publication Data

Sweet, Stephen.
Changing contours of work: Jobs and opportunities in the new economy/Stephen Sweet, Peter Meiksins.
　　p. cm. — (Sociology for a new century)
Includes bibliographical references and index.
ISBN-13: 978-1-4129-1744-5 (pbk.: alk. paper)
　　1. Technological innovations—Economic aspects—United States. 2. Labor market—United States. 3. High technology industries—United States. 4. Hours of labor—United States. 5. Industrial relations—United States. 6. United States—Economic conditions—21st century. 7. Globalization. I. Meiksins, Peter, 1953– II. Title.
HC110.T4.S88 2008
331.0973—dc22　　　　　　　　　　　　　2007031680

This book is printed on acid-free paper.

07　08　09　10　11　　10　9　8　7　6　5　4　3　2　1

Acquisitions Editor:	Ben Penner
Associate Editor:	Elise Smith
Editorial Assistant:	Nancy Scrofano
Production Editor:	Karen Wiley
Copy Editor:	Robin Gold
Proofreader:	Jenifer Kooiman
Typesetter:	C&M Digitals (P) Ltd.
Indexer:	Holly Day
Cover Designer:	Candice Harman
Marketing Manager:	Jennifer Reed

Contents

List of Exhibits

About the Authors

Stephen Sweet is an assistant professor of sociology at Ithaca College. His studies of work, and its impact on and off the job, have appeared in a variety of publications—including *Sex Roles, Research in the Sociology of Work, Family Relations, New Directions in Life Course Research, Journal of Vocational Behavior, Journal of Marriage and the Family, Innovative Higher Education, Teaching Sociology, Generations,* and *Community, Work, and Family.* His books, *College and Society: An Introduction to the Sociological Imagination* and *Data Analysis with SPSS: A First Course in Applied Statistics* (Allyn & Bacon 2001, 2003) have been extensively adopted in sociology courses. Most recently, he coedited the *Work and Family Handbook: Interdisciplinary Perspectives, Methods and Approaches* (with Marcie Pitt-Catsouphes and Ellen Ernst Kossek; Erlbaum 2006). In addition to employment as a child laborer (delivering papers), and exploitation in the fast-food industry, his work history includes a 2-year stint as a professional carpenter. In his job at home, he provides support to (and receives support from) his wife Jai (a college administrator) and children Arjun and Nisha (unemployed).

Peter Meiksins is Professor of Sociology at Cleveland State University. He has published widely on the sociology of work, particularly the sociology of technical work and of the professions, in journals such as *Work and Occupations, Sociological Quarterly, Work, Employment and Society, Theory and Society, Technology and Culture,* and *Labor Studies Journal.* He has coedited several books on work and labor, and is the coauthor of two previous books: *Engineering Labour: Technical Workers in Comparative Perspective* (with Chris Smith; Verso Books, 1996) and *Putting Work in its Place: A Quiet Revolution* (with Peter Whalley; Cornell University Press, 2002). He is currently working on a new project on the sociology of design work (also with Peter Whalley), focusing on the cases of interior, graphic, and industrial design. He is married (to a historian) and has a high-school-age daughter. He does all the cooking at home (he likes to cook; his wife doesn't); his wife barbecues and fixes the car.

Preface

This book is an effort to make sense of work opportunity—as it was in the 20th century and as it is today—and how it influences lives on and off the job. When we began this project a few years ago, we thought that this would be a straightforward endeavor. First, we intended to discuss the "old economy" and the types of opportunities present when most of the labor force was employed in jobs critical to mass production industrial work. Then we were going to write about the emerging "new economy" and the ways new technologies, new organizations, new jobs, a new workforce, and globalization are transforming work. Our unique contribution would be to show *structural lags*, the ways current policies and practices, designed to correspond with needs in the old economy, fail to address the present-day concerns of working families.

We spent well over a year blocking out chapters, going back into the research literature, writing chapter drafts, restructuring our arguments, and rewriting. But we faced a recurring problem, namely, that our observations about the old economy kept intruding into what we wanted to say about the new economy, and vice-versa. Our work in that first year would have been far easier if we had recognized then what is now a central theme of this book—*the old economy has not been replaced by a new economy; the old economy is operating within the new economy.*

Once we recognized this, we realized that our thesis would have to be modified, as would the structure of our project. The story of the old and new economies is one of *common social forces* that shape the development of work opportunity. Many features of the old economy, although sometimes in new forms, are central to the dynamics of the new. Our conclusion—one of the central points of this book—is that concerns facing workers today result from the changes in contemporary workplaces (and of structural lags in adapting to those changes) *and from enduring failures* to address the problems of inequality that developed in the old economy.

In the chapters that follow, our goals are to identify the contours of work and how they have changed over time, but also how they have remained stable. Our analysis relies primarily on the research of sociologists, but also on that of labor historians and economists. Our goal is not to offer comprehensive histories of work, or to detail the experiences of all groups in the workforce, but to document the processes that shape work opportunity and how opportunities have been divided in the United States along class, gender, and racial lines. To do this, we adopted a comparative perspective, placing our analysis of opportunity and policy in the United States alongside the somewhat different realities of work in Western Europe. We also compare the experience of workers laboring today with those laboring in the mid-20th century and earlier, and we explore the American workplace in the larger context of an integrated global economy and emerging global networks of trade.

Chapter 1, "Mapping the Contours of Work," offers an introduction to the sociology of work and the unique contributions sociological analysis brings to the analysis of the changing economy. Our concern in this chapter is not so much to analyze the nature of work in the new economy, or how changes in work have happened, but rather to indicate what needs to be examined if one is to understand work, society, and social change today. To do this, we outline observations sociologists have made about the ways culture, social structures, and agency shape the opportunity to work and the careers of workers. We introduce this chapter by describing the challenges faced by four different workers laboring in the new economy. These individuals illuminate the *diversity* of workers' experiences, and how the transition to a new economy is affecting career prospects and introducing distinct strains into family lives.

Chapter 2, "How New Is the New Economy?" considers the extent to which workplaces and labor markets have changed since the mid-20th century. We offer a critical analysis of the popular view that work has been fundamentally transformed by the advent of new technologies, new organizations, and new markets. Important changes have occurred, but we suggest that some of these changes represent "new wine in old bottles" and question whether the new economy is leading to the liberation of work and the spread of affluence. Actually, much evidence supports the conclusion that, for many workers, the opposite is happening. And when recent changes are recast in terms of class divides, we reveal sobering signs that economic transformations are contributing to a divided economy, one that sustains a two-tiered division between good jobs and bad jobs, and one that is funneling substantial shares of the returns of work to a privileged elite.

Chapter 3, "Gender Chasms in the New Economy," examines the issue of gender inequalities at work. Here we revisit the fundamental question of what constitutes work, and why women's contributions to society are commonly defined as something other than "real work" or not worthy of compensation commensurate to that received by men. We also consider the extent to which gender inequalities are disappearing in the new economy and detail why many inequalities persist. We conclude this chapter by examining the approach to handling care work in the United States, how it departs from the approaches used in Western Europe, and its impact on both the quality of care and women's life chances.

Chapter 4, "Race, Ethnicity, and Work: Legacies of the Past and Problems in the Present," examines the proposition that race might be of declining significance in the new economy. We show that racial inequalities persist but that there are important differences in the ways various minority groups have responded to, and are being treated, in the new economy. We also detail the dominant reasons why racial and ethnic inequalities exist today. Because race continues to be a major policy concern, we consider two of the most pressing debates—the controversies about affirmative action programs and about the impact of immigration on opportunity structures.

Chapter 5, "Whose Jobs Are Secure?" and Chapter 6, "A Fair Day's Work? The Intensity and Scheduling of Jobs in the New Economy," consider how security and time commitments to work have changed in the new economy. We first show the ways changing work designs in the new economy are contributing to widening job insecurity. Our interest here is not just to detail the extent of risk present today, but also to show how social policies implemented in the old economy set workers up to bear the burden of risk, often at the expense of their families and careers. Chapter 6 extends this history of the present by examining trends in the time spent working and the intensity of work. Here, we discuss the question of why American workers are working more than they did in the past, more than workers in almost every other society, and in many instances more than they want to. We also consider the implications of work in a 24/7 economy and the impact nonstandard schedules have on family lives.

Chapter 7, "Reshaping the Contours of the New Economy," outlines what needs to change if work is to become a positive experience for all and work opportunities are to be distributed equitably in the new economy. Basing our recommendations on what has been done in other developed societies, we try to offer realistic goals that, if fulfilled, would create more humane workplaces. But we also acknowledge that the dehumanizing, unjust aspects of work in the new economy are unlikely to change by themselves and that positive steps must be taken to promote improvements. A variety of agents—including individuals, activist groups, unions, corporations, and government entities—will all need to play a role in reshaping work. In the end, we suggest that government intervention will be the key to bringing the expectations of employers in line with what should be expected of workers. Its level of engagement will hinge on the ability of individuals, activist groups, and unions to exert sufficient pressures.

Our hope is that this book will help readers to understand the origins of current problems confronting working people in the new economy. Beyond this, we hope this book will contribute to a much-needed dialogue about the strategies for liberating workers from poverty, from drudgery, from discrimination, from stress, and from exploitation.

Acknowledgments

This book is the result of the contributions of numerous people, from those who cut the trees, milled them into paper, drove the paper to our offices, designed our computers, filled our libraries with books (or at least electronic links to books!), fed us and our children, and heated our offices, to those who printed and delivered it to your hands. Our intellectual efforts stand on the shoulders of the giants in the field, individuals who introduced the ideas we tried to advance and to whom our thoughts are indebted. We also relied on the efforts of the numerous bureaucrats who created the data we use to outline changes in work and opportunity. Here, we can only express our appreciation to those with whom we formed close interpersonal ties.

Our colleagues and mentors—including Cynthia Duncan, Phyllis Moen, Marcie Pitt-Catsouphes, and Peter Whalley—offered valuable guidance by directing our attention to the issues that need to be addressed and what to look for. Reviewers Judith Barker, Elizabeth Callaghan, William Canak, Carol Caronna, Marc Dixon, Linda Geller-Schwartz, Heidi Gottfried, Judith Hennessy, Martin Hughes, Arne Kalleberg, Charles Koeber, Kevin Leicht, Joya Misra, Cynthia Negrey, Vincent Roscigno, Gay Seidman, and Patrick Withen provided the sharp criticism that the book needed in its formation. Hillary Gozigian offered an insightful student's-eye view of the manuscript. Students in our graduate and undergraduate classes at Ithaca College and Cleveland State University, many of whom are also experienced workers in the new economy, raised questions that stimulated our thinking for this book. We also thank the incredibly supportive team at Pine Forge Press, including Jerry Westby, Elise Smith, Robin Gold, and our editor, Benjamin Penner.

The Alfred P. Sloan Foundation provided support for the study of job insecurities (B2001–50, Stephen Sweet and Phyllis Moen, co-principal investigators). In addition to Kathleen Christensen at the Sloan Foundation, we express thanks to Yasamin Diciccio-Miller, Akshay Gupta, and the staff of the Cornell Careers Institute for their contributions to the Couples Managing Change Study. Ithaca College generously provided some release from teaching responsibilities.

Finally, we express gratitude to our spouses, Jai and Joyce, who gave us much-appreciated time to devote to this project, who listened to our struggles, and who offered their perspectives and guidance throughout.

1

Mapping the Contours of Work

Perhaps more than any other quality, the ability to plan, organize, and collectively engage in work sets human beings apart from other species. Work occupies most of our waking hours, is a crucial part of identities, and influences life chances. At the same time, work creates problems in lives and, at its worst, can become a life sentence to grinding toil in jobs that offer few intrinsic rewards and little financial compensation. Our understanding that work can liberate, but also enslave—and seeing both possibilities exemplified in the modern economy—inspired us to write this book. We wanted to take stock of work today—to consider the types of work opportunities available, chart how these jobs emerged, and gauge the impact workplace practices have on lives on and off the job. Beyond this, we wanted to reflect on how work could be organized so that it *makes sense*—so that it provides the resources people need and contributes meaning to their lives.

This chapter begins this discussion by considering "the contours of work." These contours can be thought of as the terrain on which work opportunities are distributed and traversed. The metaphor of contours is useful because, like geographic topographies, work opportunities have been etched into the landscape by long-term historical forces. Some of these forces resulted in profound changes, wherein old ways of working have been abandoned and new methods introduced. This type of radical transformation occurred in the Industrial Revolution of the early 19th century, and some argue that computer and communication technologies are having a similar effect today (e.g., Castells 2000; Piore and Sabel 1984). Other forces, however, shape opportunity landscapes in a more gradual, ongoing, and cumulative process. Gender, with its constantly evolving meanings and practices, is one such force; so is race.

1

To consider the impact of social forces on work, we introduce here a shorthand distinction that we will use throughout this book: the division between the old and new economies. This dichotomy helps us identify the very real changes that occurred in work in the latter part of the 20th century, including the introduction of computer technologies, the expansion of a global economy, shifts in the composition of the workforce, new organizational and managerial paradigms, and other changes that we will introduce in the chapters to come. The **old economy** represents the various ways of assigning and structuring work that developed in the wake of the Industrial Revolution through the mid-20th century. These features include systems that were built around mass production, gendered divisions of labor, unionized labor, and a variety of other enduring workplace practices. The concept of a **new economy** is intended to pose the question of whether the nature of work has changed, and if it has, the extent to which these changes have affected lives on and off the job. Our frame of reference throughout this book is the changing contours of work in the United States, but as they are connected to the redistribution of opportunities in the global economy.

Although we use the term "new economy," we have come to conclude that many of the present-day contours reflect the way work evolved in the old economy.[1] Those arguing that there has been a "second industrial revolution" often ignore this. There are new jobs, new workers, and new work designs, and these are changing some of the contours of the economy. But many of the features introduced by the old economy remain. These "old" features are not simply vestiges destined eventually to die out; they are thriving and may be permanent features of the new economy that will continue to develop during the 21st century. Sometimes these old and new features are combined, for example, when old work practices are moved from the developed world to the developing world. The jobs may not have changed fundamentally, but the people who are performing them have. In Chapter 2, we consider the issue of the old and new economies in greater detail and assess the extent to which the new economy has changed and the extent to which it has remained the same.

Our discussion in this chapter is directed to identifying the dominant social forces that shape work opportunity. We organize this discussion by considering three interlocking concerns:

- **Culture**—meaning systems that attach individuals to work, harness their commitments, and direct their efforts.
- **Structure**—opportunities, as well as constraints, that shape what types of jobs can be pursued, and by whom, and the returns received.
- **Agency**—people's efforts, whether as individuals or in groups, to direct their own biographies, shape the lives of others, and respond to and sometimes modify the structure and culture of work.

To open this discussion, we consider the lives of four workers laboring in the new economy, and the rewards, strains, and constraints work produces in their lives. As you read these examples of what work is like in the new economy, reflect on the ways current opportunity structures fail to provide the resources needed, and think about how work provides meaning but also disrupts lives. The challenge, we argue throughout this book, is considering the best means of bringing culture, structure, and human initiative into harmony. In other words, the goal is to reduce the incompatibilities between how work is arranged and what workers can bring to—and receive from—their work.

Scenes From the New Economy

The experiences of Eileen, Dan, Jamal, and Chi-Ying reveal how work lives on and off the job are being shaped by the contours of the new economy. All of these cases illustrate that the effects of historical change (in this case the transition to a new economy) can vary depending on its timing with respect to an individual's biography, as well as to his or her gender, class, and race (Elder 1999; Moen 2001b). Their lives are unfolding as new opportunities are being introduced and as old opportunities are being dismantled—a dynamic Matilda White Riley identified as "aging on the moving platform of history" (Riley and Riley 2000).

Exhibit 1.1   Eileen: A Mother Strives to Mesh a High-Powered Professional Career With Family Demands

---

Eileen is a busy professional engineer in a large corporation. Early in her career, she worked 10- or 11-hour days while her husband was in school and continued to work long hours after he graduated. She was a dedicated professional and was on the fast track climbing the corporate career ladder—until she had her first child. At that point she decided that she should cut back on her hours, something she hadn't originally planned to do. It wasn't that she preferred being at home to being at work (she said being home was more "boring") but she felt someone had to do it and her husband was reluctant to cut back his hours because he had less seniority.

She persuaded her boss to let her try working 60% of her normal hours on a trial basis. He reluctantly agreed. She worked this new schedule for a few months, but found that her boss still expected more output than she could give and that he was unhappy with the arrangement, even though she still was doing a good job in almost everyone else's opinion.

(Continued)

(Continued)

   When her company reorganized, Eileen was among the first to be laid off. After losing her job, she realized how much she loved her job and the extent to which being a full-time, stay-at-home mom did not suit her interests. She worked hard to find another job at the company, one that required her to return to full-time work. But life has not been easy. Her second child was born with health problems and she and her husband frequently face situations where someone has to leave work to attend to their son. Eileen would like to get another part-time arrangement, but knows that a part-time job where she works means no promotions. So, she continues to work full-time and to wonder why she has to feel bad about being away from work when her child is ill.

Based on Arlie Russell Hochschild, *The Time Bind: When Work Becomes Home and Home Becomes Work.* New York: Metropolitan Books, 1997, pp. 88–98.

Exhibit 1.2   Dan: An Insecure, Older Worker in a Declining Industry Strives to Salvage a Career

Dan had been an autoworker at GM's Linden Plant in New Jersey for about 10 years. He started working there when he was just 20, thinking it would be just for a while, and he never really enjoyed the work. But the money was good, and he found himself thinking about working his way up to a supervisor's position. While he was working at GM, Dan started a chimney sweep business on the side. He started out small and continued working at both jobs, especially after his wife had kids and quit her job. After a few more years of working two jobs, Dan realized that his GM job was in jeopardy; the union wasn't as strong as it had been and there were rumors of cutbacks and layoffs. He stayed, though, since it didn't seem that layoffs would happen tomorrow and his paycheck was important. Then, the company offered workers a buyout. Dan decided that, rather than risk losing his job *and* the buyout, he'd take advantage of the offer.
   He used the buyout money to build up his chimney sweep business, buy new equipment, etc. He did really well at first, and initially felt he had made the right decision—he liked the work better and was making a good living. But then the economy took a turn for the worse, and his business struggled. He hung on by scrambling to find other odd jobs. He likes what he's doing, but he wonders whether he made the right decision; he knows staying at GM meant doing a job he didn't like, and he would have continued to worry about layoffs, but he also knows that his business could fail and that he'd have to start over again.

Based on Ruth Milkman, *Farewell to the Factory: Auto Workers in the Late Twentieth Century.* Berkeley: University of California Press, 1997, pp. 3–6.

Exhibit 1.3   Jamal: A Disadvantaged Young Worker Strives to Start a Career

Jamal is a 22-year-old African American man living in Harlem with his common-law wife. They live in a tiny, untidy one-room apartment where they have little more than heat and a place to sleep. Jamal didn't finish high school and eventually got his GED. He is intelligent and ambitious, but was not a dedicated student. He has worked since he was very young, partly because he was determined not to become like his mother, who had a drug addiction.

Despite putting a great deal of effort into finding work, he had very little luck finding a good job. The best position he was able to find was at a fast-food restaurant an hour's commute away, which he found through some of his friends who also work there. The job doesn't pay well (minimum wage) and he often is not able to secure the full 8-hour shift he needs to get by. His boss often sends him home after 5 hours, but sometimes expects him to work longer if needed, or less if business is slack.

Jamal and his wife do not get help from their parents. His mother-in-law is furious with her daughter for marrying Jamal, his father was never in his life, and his mother is a drug addict. He is pessimistic about his future, and expects to continue working in jobs like the one he has now. But he keeps working, hoping that maybe he will be one of the lucky few who gets a job in a car factory.

Based on Katherine Newman, *No Shame in My Game: The Working Poor in the Inner City*. New York: Alfred A. Knopf and the Russell Sage Foundation, 1999, pp. 3–13.

**Exhibit 1.4   Chi-Ying: A Daughter Strives to Carve a Career in an
                 Industrializing Economy**

Chi-Ying is a young peasant girl from a rural village in Northern China. Two years ago, she migrated to Shenzen in the South to work at an electronics plant. The work is hard and unpleasant, sometimes exposing her to hazardous fumes. She is required to work 11-hour shifts and sometimes has to work overtime as well (or she might be fired). Being absent leads to fines and punishment, so she comes to work even if she's ill. Her apartment is cramped, she lacks access to clean drinking water, and the air that she breathes is often a choking smog. This is the result of both lax regulation and the proliferation of highly polluting industries, which have transformed Shenzen from a rural countryside into an urban center within a few short decades.

Chi-Ying's move to Shenzen was supposed to be temporary. Her parents actually arranged a marriage for her a year ago, but that fell apart when she said she wanted to continue working for a few more years. She had to use some of her own wages to pay the groom's parents back for presents and gifts her family had received. Chi-Ying still wants to keep working at Shenzen, despite the hardships. She has her own income, which amounts to more than half of what her father earns back home as a farmer.

(Continued)

(Continued)

Although she doesn't have much time for leisure, she does have the opportunity to buy a few things for herself and to experience life outside her village. She has married a boy she met in Shenzen (who comes from her part of the country), and she knows she eventually will return. But she is pleased that, unlike her grandmother, she has seen life in the city and has had a job of her own.

Based on Ching Kwan Lee, *Gender and the South China Miracle*. Berkeley: University of California Press, 1998, pp. 3–9.

For Eileen and Chi-Ying, economic changes have opened new opportunities, but they coexist with enduring sets of expectations about what mothers and daughters should provide to their spouses, parents, and children. These cultural orientations seem more appropriate to another era and lag behind what these workers ideally want for themselves and, in many instances, what they can provide for others. For Dan, the moving platform of history has introduced new career tensions because the types of jobs he performed during his entire career are becoming more difficult to find. Consequently, his income is in jeopardy, and he faces the daunting challenge of fitting himself into new lines of work that he has little experience performing (Sweet 2007). But because he was born in an earlier generation (and possibly also because he is of a different race), he has had opportunities not available to Jamal. Jamal is doing the "right thing," working hard and trying to get ahead, but his opportunities are limited by the fact that good well-paid jobs that only require a limited education are disappearing in the new economy.

The careers of these workers are influenced by demands and social ties off the job. All these workers are making career decisions in the context of their linkages to others. In some circumstances, parents hold sway, whereas in other cases, it is the needs of spouses, children, or both (Neal and Hammer 2006; Sweet and Moen 2006). These life stage circumstances play an important role in shaping worker behavior, expectations, and needs. How people respond to these circumstances is heavily influenced by cultural scripts (e.g., assumptions about what parents should provide for their children) and the availability of resources (which varies from person to person, group to group). And beyond family ties, the context of neighborhoods and communities influences one's ability to find work, the resources to prepare for work, and the security to engage in work (Bookman 2004; Sampson, Morenoff, and Gannon-Rowley 2002; Sweet, Swisher, and Moen 2006; Swisher, Sweet, and Moen 2005; Voydanoff 2007).

Culture and Work

Eileen presents an interesting case to consider because she is a worker who could potentially leave her job to tend to her children. She feels guilt about not doing so, but chooses to stay in the labor force and retains ambivalent feelings about her choices. Understanding why workers like Eileen place such a strong emphasis on their work roles requires considering culture—the meaning system that surrounds work and shapes identities in respect to it.

Most classic theories of work embrace cultural perspectives that view labor, in and of itself, as a noble endeavor. Karl Marx (1964 [1844]), for example, argued that work is what distinguishes humans from other species, and highlighted how it enables people to transform their environment to suit human interests. Sigmund Freud (1961 [1929]) argued that work is a socially accepted means by which humans are able to direct their sublimated sexual energies. As such, he saw work as a means of achieving satisfaction when fulfillment in other parts of life is lacking or prohibited. Émile Durkheim (1964 [1895]) offered a different thesis, that work and the complex division of labor in society offered a means to create social cohesion. All these perspectives have in common the assumption that work has the potential to cement social bonds and advance the development of civilization.

But has work always been embraced by cultures as being a central role in people's lives and the workings of societies? Anthropological and historical studies suggest otherwise. In many cultures, work is defined as the means for day-to-day survival. Subsistence economies operate on the basis of cultural assumptions that work is primarily a means to an end, so that once individuals have enough food and shelter, labor is expected to cease. Such an orientation to work in today's American culture would indicate a moral weakness and be perceived as a threat to social order. But from the point of view of many other cultures, our embrace of work could be considered pathological. If one can obtain enough to eat and gain sufficient shelter by working a few hours a day, so be it. Why should a hunter set out in search of game if the supply of food is adequate (Brody 2002; Sahlins 1972)?[2]

One important cultural question concerns why work plays such a central role in some societies but not in others. Part of the answer, according to Max Weber (1998 [1905]), is that the societies that were in the forefront of the industrial revolution had been swayed by changing religious doctrines. These religious beliefs, particularly those that underpinned the Protestant Reformation, created anxieties about one's fate in the afterlife. In response, Western European and American culture advanced the value of the **work ethic,** a belief that work is not something people simply do, but is a God-given purpose in life. Devoting oneself to work and doing a good job were considered to be ways of demonstrating to oneself that a life of virtue reflects grace. And as members of these societies embraced the idea that work is "a calling," they applied themselves to their jobs with greater vigor, creating wealth and affirming to themselves and others that God was looking favorably on their actions.

Although many now question Weber's thesis that the Protestant Reformation was responsible for the emergence of capitalism, the centrality of the work ethic to the development of Western society is widely accepted. So deeply is it ingrained in contemporary American culture that nearly three-quarters of Americans report that they would continue to work, *even if they had enough money to live as comfortably as they would like for the rest of their lives*.[3] Americans work to affirm to themselves and others that they are virtuous, moral individuals, good people who deserve respect (Shih 2004). Conversely, those who choose not to work, or workers like Jamal who are unsuccessful in securing a job, are looked down upon and stigmatized. In American society, to be without work is to be socially suspect and unworthy of trust (Katz 1996; Liebow 1967).

Although the work ethic defines labor as a virtue, it also has pathological dimensions. The cultural embrace of work may be akin to the flame that attracts the moth. It is telling that many who can afford to work less, and who have the opportunities to do so, choose not to (Hochschild 1997). Psychologists call these individuals "workaholics," (Machlowitz 1980), but as we discuss later in this book, many of those driven to work long hours do so because they are driven by organizational cultures that bestow rewards on those who live, breathe, and eat their jobs. The suspicion cast upon those who do not hold jobs has created pressures to force work upon those who get little benefit from it. Consider that welfare reform legislation, passed in the mid-1990s, requires even very poor mothers of young children to work to receive welfare assistance. This requirement defines mothering as "not work" (a concept we return to later) and accepts the fact that many of the affected mothers remain in poverty even after they are employed.

Thorstein Veblen (1994 [1899]) in *The Theory of the Leisure Class* observed that attitudes to work are bound up with materialistic values held in American culture. Markers of status include luxury autos, large homes, and expensive clothing. All of these markers of success are conspicuously consumed, put on display to be seen and admired, and set standards for others to follow. By the mid-20th century, the drive to purchase social status had permeated American society, compelling workers to labor hard "to keep up with the Joneses" and their neighbors' latest purchases (Riesman, Glazer, and Denney 2001 [1961]). Contemporary American workers engage in the same status game that emerged in the late 19th century, but with new commodities (i.e., iPods, BMWs, and flat screen TVs). Their competition now expands beyond their neighborhoods, as they are literally saturated with media images of success and have developed numerous ways to accumulate debt (home equity loans, student loans, and credit cards) (Gergen 1991).[4] The result, some have argued, is "affluenza," the compulsion to purchase and spend beyond one's means (Graff, Wann, and Naylor 2001).

For some members of the new economy, work has become the means to manage spiraling debts incurred while striving to keep up with others who are *also* spending beyond their means (Schor 1998).

Culture also shapes the attitudes workers and employers have toward each other. One means by which it does this is by constructing social divisions and setting group boundaries. Racial and gendered divisions, for example, are based on assumptions that different social groups possess different capabilities. In turn, these beliefs contribute to the formation of self-fulfilling prophecies. Whether or not these differences were originally real is immaterial; as the early 20th century American sociologist W. I. Thomas noted, what people *believe* is real often *becomes* real in its consequences (Thomas and Thomas 1928). As we discuss later in this book, these self-fulfilling prophecies about gender and race shape social networks, influence access to resources, and funnel people into different lines of work.

Culture even extends into the design and management of jobs and technologies. Consider, for example, the enduring legacy of **scientific management.** Frederick Winslow Taylor introduced this managerial philosophy (also known as Taylorism) at the beginning of the 20th century to increase the productivity of workers laboring in factories. He advocated the benefits of redesigning work to wrest control from workers and place it in the hands of management. His *Principles of Scientific Management* (1964 [1911]) argued for the separation of "thought" from "execution" to establish clear divisions between managers (whose job was to think and design) and workers (whose job was to carry out managers' instructions). He used time-motion studies to decompose production jobs into the simplest component tasks in order to increase worker speed and accuracy. And managers' jobs were redefined to absorb worker skills into the machines and organization and to keep the flow of knowledge going in one direction—from the shop floor into managers' hands. The result was the creation of legions of deskilled jobs, the dissolution of many craft skills, and a decline in the individual worker's ability to control the conditions and rewards of work (Braverman 1974; Noble 1979; Pietrykowski 1999). It also fostered distrust and hostility between workers and their bosses (Montgomery 1979).

Why did Taylor advocate this way of organizing work, given its obvious negative consequences for the quality of work life and its negative effects on labor–management relations? In part, it was a response to something real—the fact that workers often *did not* work as hard as they could. His experiences had taught him that they did not show up to work consistently, took long breaks, and worked at a more leisurely pace than owners desired. His interpretation of this behavior, however, was culture-bound. Taylor interpreted workers' behavior not as a rational, class-based resistance to employers but as an *irrational* unwillingness to work in the right way.

Taylor, like many Americans of his time, was embracing a cultural denial that class divisions within the workplace existed. His solutions also reflected the culture in which he was living. He advocated a reorganization of the workplace based on scientific methods, something that resonated tremendously in a society where science had come to be seen as the solution to many human problems. And he depicted the worker as essentially unintelligent and easily manipulated; Taylor was fond of using an example involving a worker named Schmidt (whom he described as "oxlike"), whom he persuaded to adopt his new system through a combination of simple-minded arguments and limited incentives. This, too, was typical of American culture at that time; many Americans believed that members of the lower classes, immigrants, and others at the bottom of American society were inferior in various ways (including intelligence) to the more successful members of society. Taylor's ideas also reflected an abiding cultural belief in the correctness of **capitalism**, particularly the proposition that it is natural that some should be owners and others laborers, that the efforts of those at the top were more important and valuable, and that an extremely

Exhibit 1.5   The Film *Modern Times* Offered a Poignant Illustration of the Alienating Nature of Work in Factory Jobs in the Old Economy

Source: © Getty Images. Reprinted with permission.

unequal distribution of the fruits of labor was not just defensible but actually desirable (Callahan 1962; Nelson 1980).

The legacy of managerial philosophies—in this case, scientific management—highlights how culture and social structure intersect. Managerial perspectives that embraced the proposition that workers are indolent and should not be trusted are directly responsible for the creation of many of the alienating low-wage "McJobs" present in America today. These philosophies initiated the development and application of assembly lines, promoted the acceptance of the idea that some people should be paid to think and others to labor, and fostered divisions between "white-collar" and "blue-collar" jobs. Dan performed blue-collar work during the bulk of his career, and these approaches to organizing work are directly responsible for shaping Jamal's tasks in the fast-food restaurant.

Culture, then, is an important force shaping the contours of work. These examples of how culture shaped workplaces in the past suggest interesting questions about culture's role in carving out the contours of the new economy. Have cultural attitudes about the role of work changed, and if so, have workplaces changed along with them? How long are people working and why do they work so much? Have Americans begun to abandon long-standing (Taylorist) cultural assumptions about the proper way to organize work, or do we continue to construct workplaces on the assumption that workers are lazy, ignorant, and not to be trusted? To what extent are perceived divisions between the members of society continuing to deprive some people of access to opportunity?

Structure and Work

Although culture creates meaning systems that orient people to work, social structures involve enduring patterns of social organization that determine what kinds of jobs are available, who gets which jobs, how earnings are distributed, how organizational rules are structured, and how laws are formulated. Social structure does not exist independently of culture. Often, social structure reflects cultural attitudes (because people tend to create institutions that are consistent with their beliefs), but it also can be in conflict with aspects of culture, creating tensions and contradictions with which individuals and societies must grapple (consider how the structural reality of unemployment creates particularly difficult problems in a society where work is "mandatory"). Throughout this book, we will discuss various aspects of social structure, the access to different types of work, the division of labor, the social organization of workplaces, and legal and political arrangements that structure workplaces. Here, we simply illustrate how social structures affect individuals' experience of work by considering how a few aspects of social structure—social class, job markets, and labor force demographics— may be affecting opportunities and workplace practices in the new economy.

Class Structures

Witnessing the changes wrought by industrialization, Marx (1970 [1867]) focused sociological analysis of work on **class structures,** the socio-economic divisions between different segments of the workforce. In his classic analysis of the industrial capitalist economy, he argued that employers' profits depend on the effort put forth by employees, which created incentives to limit wages and to push workers to labor as hard as possible. He also observed that the efforts of workers created far greater wealth for employers than it did for employees. Considering these class relations, Marx argued that the tendency for work under capitalism would be toward the creation of a polarized class structure, comprising a disenfranchised working class (the proletariat) and an affluent owner class (the bourgeoisie).

The class structure of the United States is more complex than the polarized structure Marx described. Although workers and capitalists exist, large portions of the workforce seem to fit into neither category. For example, numerous professional and managerial workers have substantial education, some (or even considerable) workplace authority, and higher salaries than the typical front-line worker. Yet, it is difficult to describe them as captains of industry or members of the dominant class, given that they are not in charge and work for someone else (who has the ability to fire them). Sociologists have argued long and hard about how to describe these intermediate class positions. One neo-Marxist sociologist described such workers as occupying "contradictory class locations," combining elements of the classes above and below them (Wright 1985).

Although the precise shape of the class structure of capitalist societies is a matter for dispute, what is not disputed is that class matters. For example, class affects people's access to work opportunities, although the precise way in which it does so has changed over time. Before the Industrial Revolution, most children inherited their line of work from their parents through a process known as **ascription.** Farmers' children tended to become farmers themselves, and craft workers would often learn their trade from their fathers. Women's roles were largely ascribed as well. One's occupation, then, was to a great extent one of the things one inherited from one's parents; the cross-generational effects of class were obvious and straightforward. With industrialization, however, the range of jobs expanded profoundly, many new occupations were introduced, and other occupations became less common or actually disappeared. As a result of these changing opportunity structures, fewer children could follow in their parents' footsteps or inherit occupations from the previous generation. By the late 19th century, geographic mobility (and social mobility) was common, as children ventured further from their home communities to find work (Thernstrom 1980). Class still mattered, however, because it affected one's access to resources such as education, skills, and connections that determined access to jobs in an economy where jobs no longer were inherited.

The existence of class also affects the structure of workplaces. Marx felt that the antagonism between labor and capital inevitably produced antagonism at work and led to the development of hierarchical, top-down managerial structures designed to control workers and ensure that the interests of employers predominated. Although the polarized workplaces envisaged by Marx may not be the dominant organizational form, Taylorist ideas about the need to control labor reflect a strong desire to respond to class antagonisms. They reflect the reality that the workplace is a zone of contested terrain, one in which class conflicts take place, with each side using the weapons at its disposal—including layoffs, speedups, technology, strikes, and even sabotage (Edwards 1979; Montgomery 1979).

Throughout this book, we argue that social class remains one of the most powerful forces shaping employment opportunities and access to resources in the new economy. We examine how a changing economy has altered the reality of class and the extent to which changes in class structure have led to a fundamental restructuring of workplaces away from the familiar patterns of industrial America. We will also examine how gender and race matter and how they interact with class to shape complex, contemporary structures of opportunity and workplaces.

Job Markets and Job Demands

Jamal's and Dan's problems involve not so much finding work as finding work that pays a reasonable income. Eileen, on the other hand, possesses highly marketable skills and can command a handsome salary. For her, the problem is securing a job that is designed to correspond with what she can bring to the job. All three of these workers have concerns that are structural in nature and involve the way opportunities are configured. All have to adapt themselves to the existing range of jobs and the prevailing ways in which jobs are organized. Their personal problems reflect the fact that workers—especially those laboring in times of economic change—face challenges in locating and adapting themselves to opportunities.

The Industrial Revolution of the early 19th century was clearly a watershed, one that profoundly reshaped the types of jobs available to workers. The most obvious consequence of industrialization was that far fewer people were employed in agriculture and many more were employed in factory work. However, the changes were not limited to the shift from agriculture to industry. Traditional occupations outside agriculture were also transformed, as new technologies and new ways of organizing work pushed older approaches aside. Weaving, for example, was once a task performed in the home. The mechanization of weaving during the Industrial Revolution, however, completely eliminated this form of work and transformed the skills into those fitting factory labor. Similar stories can be told about many other traditional occupations, including hat making, shoe production, tanning, and tinsmithing (Thompson 1963).

Is the range of employment opportunities available to American workers changing again? It certainly seems that way. Some jobs that used to be plentiful in America have virtually disappeared, and the skills needed to obtain jobs are changing as well. In the old economy, for example, it was common for children to follow their parents into the mill or factory and receive good wages for performing jobs that required little education. But today, few young people aspire to become steelworkers or factory operatives, largely because many of these jobs have disappeared. As steel mills and factories closed in the 1970s and 1980s, the impact reverberated throughout industry-dependent "rust belt" communities, forcing their residents to rethink long-standing beliefs about jobs, futures, and how one makes a living (Bartlett and Steele 1992; Bluestone and Harrison 1982; Buss and Redburn 1987).

One way of considering the changing opportunity landscape is to consider the process of **creative destruction,** a phrase introduced by the economist Joseph Schumpeter (1989) to describe the tendency for old methods of production to be replaced by newer, more efficient approaches. In some cases, new technologies make old needs obsolete, as when the automobile extinguished the need for buggy whips. In other instances, technological innovation can replace workers with machines, as was the case with cigarette rollers (Bell 1973). New methods of organizing work can also be used to reduce production costs, for instance, by moving jobs to locales where labor costs are lower (Cowie 2001). And in the case of computers, technologies have not only replaced workers, but also introduced entirely new markets and jobs.

The drive to create ever more efficient and profitable enterprises is influencing the distribution of work opportunities around the world. Production now occurs on a global scale, and the forces that disperse work to far-flung locations such as Indonesia (where athletic shoes are assembled) and Vietnam (clothing) shape the life chances of workers both at home and abroad. Understanding the reasons why work is being dispersed, and the impact on workers' lives at home and abroad, is essential to revealing the trajectory of work and opportunity in the new economy. Throughout the 20th century, the

United States held a dominant position in the global economy. But in the new economy, jobs previously held by Americans such as Dan are increasingly being exported to countries like China and India and being performed by workers like Chi-Ying. Is this resulting in deteriorated or enhanced opportunities? And for which workers?

Changing employment opportunities also have redefined what skills are needed, reshaped job demands, and introduced new rewards. They also take new tolls on workers' lives. Consider the large number of jobs available in various kinds of **interactive service work** that emerged in the latter part of the 20th century. These jobs require a different type of work than that performed in the factory, in that the employees typically do not manufacture anything. Their jobs (such as teacher, therapist, or server) involve providing a service for someone else with whom they are in direct contact. Sociologists have noted that this kind of work places different demands on the worker (Mills 2002 [1951]; Paules 1991). He or she must learn interaction skills—how to make others feel comfortable, how to produce the desired kind of social setting, how to deal with various kinds of difficult social situations—*because the interaction is a significant part of the product being sold.* The work of airline flight attendants offers a compelling illustration because they are trained to make customers feel safe and at home in the rigid and sometimes frightening environment of an airplane. To do this, these workers are coached on techniques to change their internal emotional states to generate the display of warmth or sex appeal required by their employers. As a consequence, however, these types of workers are especially prone to experiencing emotional numbness or burnout (Hochschild 1983).

Although new jobs demand new sets of skills, new technologies and organizational systems are also transforming many familiar jobs. A secretary's job, for example, is quite different than it once was because computers have eliminated aspects of the old job (repetitive typing) and created new ones (basic graphic design, data analysis, electronic communication). Bank tellers once were simply clerical workers who processed clients' financial transactions. Now, however, computerized information systems provide tellers with information about clients' financial positions and prompt tellers to sell various products to the client, all while maintaining a close electronic eye on what the worker is doing (Smith 1990). Even traditional manual labor is affected. For example, production workers who used to rely on their senses of touch and smell as guides now work in clean settings and operate sophisticated computerized systems that make some of their old ways of working obsolete (Noble 1979; Shaiken 1984; Vallas and Beck 1996; Zuboff 1988).

Finally, job opportunities may be less rigidly tied to space and time than they were in the old economy. Today, many workers have opportunities to telecommute and work from home offices. The economy operates 24/7, introducing the prospects of working alternate shifts and reconfiguring work around family lives. This may open opportunities to liberate workers from the traditional 9 to 5 grind and introduce new flexible schedules that more harmoniously mesh work with life—a work arrangement that Eileen's boss was reluctant to accommodate. However, it may also open prospects that work will intrude on lives in ways not possible in the old economy. Understanding the impact of these new structural configurations is essential to charting the contours of work in the new economy.

Demography and the New Labor Force

The composition of the workforce is also undergoing change. As a result, sociological analysis of work requires consideration of **demography** and of how the composition of a society affects the placement of workers into jobs and the distribution of opportunities to prepare for and obtain work (Farnesworth-Riche 2006). The paid labor force is quite different today than it was in the mid-20th century or earlier. It contains a far higher percentage of women, and its racial and ethnic makeup is different. We devote two chapters of this book specifically to the issues of gender and race/ethnicity; here we introduce the importance of demographic forces by considering how age structures affect the availability of jobs, the availability of workers, the need to work, and the returns received from work.

The U.S. labor force, along with those of many other developed societies, is aging. Americans today can expect to live 12 years longer than could those alive in 1940, and 26 years longer than those who were alive in 1900.[5] Workers are living longer, and they are healthier when they reach ages that used to be considered "old." This presents new opportunities, as well as new challenges, to American workers and their employers, such as the approach to dealing with retirement. Should workers continue to stop working at 65 if they are going to live for many years after that? If people are living longer and staying healthy longer, perhaps work careers should be lengthened. However, older workers generally do not want jobs that demand heavy schedules. More common are desires to enter into second or third careers and to pursue work situations that focus less on earning money (although for many that remains important), and more on satisfying creative desires or making a difference in the lives of others. Unfortunately, most employers do not offer "bridge jobs" that accommodate the possibility of the types of scaled-back employment that fit the skills and interests of these workers (Hutchens and Dentinger 2003; Moen and Sweet 2004; Moen, Sweet, and Swisher 2005).

Exhibit 1.6 Age Distributions in the United States: 1940 and 2000

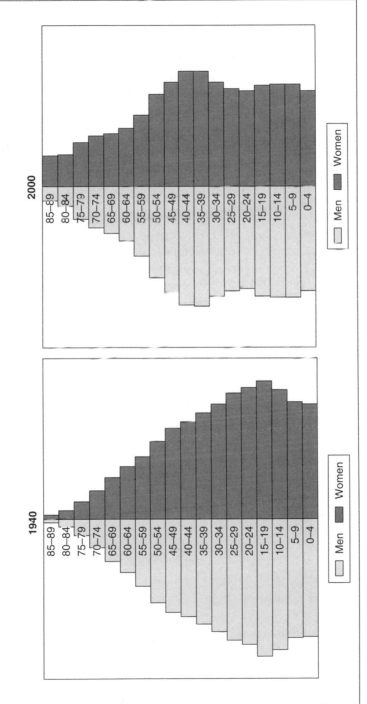

Source: Statistical Abstracts of the United States

The changing age structure of the workforce presents challenges to society as a whole, not just to employers. Exhibit 1.6 shows how the age structure of the United States has changed from 1940 to 2000. Note that in 1940 the age structure of the United States resembled a pyramid, with most of the population in the younger age groups, with a steady attrition as one approached old age. Only a relatively small group lived beyond age 70. In contrast, in 2000 the age pyramid looks more like a skyscraper, albeit with a bulge in the middle. This bulge is the baby boom generation, a birth cohort that is steadily aging its way into retirement years. A key structural question concerns how an aging society will provide economic support for the growing numbers of older people. Will they be required to work? Or will society continue to provide post-employment pensions for them? And, if the latter, how will that expense be financed? The Social Security system, the most important source of retirement income for many Americans, is funded through taxes on currently employed workers. Those taxes become part of the general pool of Social Security revenue, which provides pensions to those who have retired. Some policymakers are concerned that if the pool of retired workers becomes larger and the pool of employed workers becomes smaller, the revenues available to fund the system will be squeezed (Weller and Wolff 2005). There is much controversy about whether this should be called a "crisis," but there is general agreement that ways need to be found to ensure that adequate revenues will be available for the growing population of retired workers.

Demographic factors such as age, gender, and race affect virtually all aspects of the economy and workplace. Demographics play a role at the organizational level, as the experiences of ethnic minorities and women are commonly shaped by their scarcity at the top levels of organizational hierarchies. They are critically important at the community level, as neighborhoods that lack job opportunities hinder the socialization of children into the types of workers needed in the new economy. We will return to the critical issues of aging, gender, race, education, and immigration throughout this book.

Agency and Careers

Sociologists are often accused of arguing that people are simply "pawns" or "cultural dopes" of the larger social structural and cultural contexts in which their lives are lived. The depiction of individuals as victims of external forces ignores **agentic capacities**—their ability to direct their own lives and those of others (Garfinkel 1967; Wrong 1961). All of the workers we considered made choices. Jamal got married at a young age and dropped out of high school, Eileen elected to have two kids and pursue a high-powered career, Dan took the initiative to start his own business, and Chi-Ying chose to move from her village to the city. These observations highlight the ways in which different people direct their life courses and how access to different resources and constraints shape how lives are constructed over time (Elder 1998; Moen 2001a; Sweet and Moen 2006). The life course perspective is essential to understanding the contours of the new economy because it focuses on **careers**—the patterns of entry, exit, and movement between jobs.

Agency, of course, depends partly on resources. People with unlimited resources at their disposal are in a far better position to design their own lives than are those who have few resources. The new economy may be creating a context that is expanding the control individuals have to direct their life courses, in essence making lives less scripted than in the old economy (MacMillan 2005). Many old structural barriers have been removed (such as segregation laws), and so have the cultural barriers that funneled women and ethnic minorities into restricted ranges of occupations. Before the enactment of civil rights legislation and the women's movement, the prospects of a woman like Condoleeza Rice moving into a position of power were slim to nil. Today, one can quickly generate a sizable list of minority group members and women who have moved into professions in which they had been entirely absent. Still, there is ample evidence to indicate that women and minorities are at distinct disadvantages in securing many types of jobs (Grusky and Charles 2004; Reskin, McBrier, and Kmec 1999). Whether the new economy is fundamentally altering the possibilities for people to shape their own biographies is one of the central questions posed in this book.

Agency also plays a critical role in shaping the way work is performed. Numerous ethnographic studies reveal that workers are not simply passive recipients of culture and structure; they use personal initiative to influence how their jobs are performed and the returns they receive from work (Darrah 2005; Montgomery 1979; Richardson 2006; Roy 1955; Tulin 1984). To illustrate agency at work, consider Michael Burawoy's (1979) observations of production workers in the machining industry. These workers' jobs were regulated by quotas, wherein they had to make a specified number of parts to earn their base pay. But when they surpassed those quotas, they could "make out" and earn additional money. In one respect, this system was rigged by management to increase productivity. However, Burawoy observed that the machinists invented a variety of tricks to game this system. For example, they would keep quiet about the easy jobs in which quotas were underestimated, and complain incessantly about the impossibility of meeting quotas on virtually all other jobs. They would bribe supervisors to get the easiest jobs and curry favor with coworkers to provide parts stock needed to get their jobs rolling. When given an easy quota, workers overproduced and then hid their "kitties" that could be turned in for extra compensation at a later date. In sum, these machinists showed that when workers are confronted by cultural and structural arrangements, they also engage in **strategic action** to influence how these arrangements affect their lives (Moen and Wethington 1992; Sweet and Moen 2006).

Finally, it should be added that agency also operates at a collective level. Workers make efforts to carve out work lives for themselves, but they also collaborate with others to reshape the contours of work and create more satisfactory work opportunities for themselves. An obvious example is that workers band together in organizations such as unions or professional associations that use the strength of numbers to press for needed changes. Union publicity materials that describe unions as the "people who brought you the week-end" remind us that collective action obtained the taken-for-granted days off workers now enjoy. Similarly, the professional associations formed by doctors, lawyers, and others help protect those workers from competition, define what are acceptable (and unacceptable) professional practices, and generally shape the conditions under which those types of work are performed. Throughout this book, and particularly in the concluding chapter, we will examine how collective action has shaped workplaces in the past and how it might do so in the future. Is the new economy making certain forms of collective action by workers obsolete? Is it creating openings and needs for new kinds of collective action? What are the key issues around which workers are banding together to effect change?

Conclusion

In this first chapter, we focused on the ways sociological perspectives reshape the consideration of work. Although work is commonly considered a means to obtain a paycheck, we argued that it is much more than that. The design of work corresponds with cultural templates that guide workers to their jobs and script social roles. Workers live within social structures that both allocate opportunities and construct social inequalities of access to meaningful employment. And within these contexts, workers have responded both individually and collectively to manage their responsibilities and reshape society.

The stresses experienced by workers like Eileen, Chi-Ying, Dan, and Jamal are probably familiar to many readers of this book. Because of the instability of jobs, changing opportunity structures, the challenges of meshing work with family, and the challenges of finding good work, many workers find themselves struggling in the new economy. One of the great contributions of sociology is its capacity to reframe these types of personal problems as being public issues (Mills 1959). In the chapters that follow, we will consider the extent to which work opportunities are changing, and the impact these changes are having on lives on and off the job. Our focus, throughout, is on considering stress points and gaps and how workers adapt to these stresses and opportunity divides, as well as what can be done to close the chasms that separate workers from fulfilling jobs held on reasonable terms.

Notes

1. Of course, these are not the only phrases used. Others use the term "Fordism" to describe the old economy, and depending on the political slant of the analysis, "post-Fordism" and "flexible specialization" are used to describe the new economy, as are "knowledge economy," "global economy," and "post-industrial economy" (Bell 1973; Hirst and Zeitlin 1991; Piore and Sabel 1984).

2. It is worth emphasizing that describing societies such as these as "poor" is misleading. Although they lack the variety of possessions contemporary Americans enjoy, their members often live healthy and fulfilling lives.

3. Authors' analysis of the General Social Surveys. Retrieved from http://webapp.icpsr.umich.edu/GSS/

4. Approximately one-third of American families rent their homes, one-quarter live at or near the poverty level, and nearly one-half will experience divorce. These facts are seldom represented in television's portrayals of the "typical" American family. Stephanie Coontz, 1992, *The Way We Never Were: American Families and the Nostalgia Trap*, New York: Basic Books.

5. American men now live, on average, to be 75 years old, and American women have a life expectancy of 80 years.

2

How New Is the New Economy?

One of the most popular themes in discussions of work is the idea that America is situated in a "new economy" and that recent changes in work constitute the equivalent of a second Industrial Revolution. Consider, for example, the impact computers have had on the ways jobs are performed and designed. Computers enable workers to correspond at great distances, telecommute from home, and access a wide array of information. These "smart machines" have absorbed many workers' jobs and replaced human hands with robotic pincers that move with exacting precision. Computers have also spawned new markets for software and hardware, creating new jobs requiring new skills. Their reach spans the world, enabling near-instantaneous transmission of information, as well as the coordination of complex trade relationships that link companies with one another in global webs. It is hard not to conclude that computers have sparked revolutionary changes— not only in what is being produced and how jobs are designed, but also in the geographic distribution of work.

Information technology is but one change that occurred in the context of other dramatic transformations, including the large-scale entry of women into the paid labor force, the expansion of service jobs, declines in manufacturing job opportunities, new managerial paradigms, and the push toward integrating work in America with global supply and distribution chains. What impact will these changes have on the current and future generations of workers? Do they indicate that America is on a path toward expanded opportunity, or will the well-paying jobs disappear? Can workers expect to labor in jobs that offer challenges, interesting work, and opportunities to grow, or are greater proportions of workers likely to find themselves trapped in monotonous, mindless jobs? And ultimately, if we are in a new economy, does this represent a sharp break from the past, with a completely reconfigured opportunity structure, or does it indicate more of the same—modifications of the general trends and tendencies that emerged in the old economy?

The use of the concept of a "new economy" (or alternate terms such as "global economy") is widely accepted as a shorthand way of saying that work today is remarkably different than it was in the recent past. But in this chapter, we open this assumption to debate. If there is a new economy, what are its distinguishing characteristics? We argue that job opportunities *have* changed in profound ways. There are new technologies, organizational designs, industries, and markets. The economy has become increasingly international. These changes have introduced the need to develop new skills to fit changing opportunity structures. But what is equally true is that many aspects of the "old economy," including the design of jobs to require limited skill, have either survived or been reproduced in new forms. After all, for every successful computer programmer who works at companies like Microsoft, one can find three poorly paid workers laboring on hamburger assembly lines in companies like McDonald's.[1] Understanding the new, the old, and the old in the new is the key to understanding the diverse needs and experiences of today's workforce.

The Old in the New

In the sections to follow, we consider some of the major changes said to characterize work in the new economy, including the decline of mass production and manufacturing work, the emergence of new cultures of control, the gradual disappearance of organized labor, and globalization. In each case, we argue that there *have* been significant changes, but also that there are persistent features that reflect the perpetuation of the old economy within the new.

A Post-Industrial Society?

One of the earliest forecasts of an emergent new economy came from the sociologist Daniel Bell (1973), who argued in the early 1970s that America was entering a "post-industrial" era, in which the manufacturing-centered economy of the past was being replaced by an economy directed toward the provision of services. Bell was among the first to note something that subsequently became obvious to most Americans, particularly those located in the so-called rust belt of the industrial Midwest—employment opportunities had shifted away from manufacturing to other sectors of the economy.

Exhibit 2.1 shows that in 1939 the number of employees working in the manufacturing sector in America equaled the number of employees in every other sector of the economy *combined*. Until 1989, the manufacturing sector remained the dominant employer. But as the population of the United States grew during the latter part of the 20th century, manufacturing employment did not. Today, instead of employing 1 in 2 workers, as it did in the mid-20th century, manufacturing enterprises only employ 1 in 10 workers.

Exhibit 2.1   Trends in Employment in 12 Major Sectors: United States
1939–2005

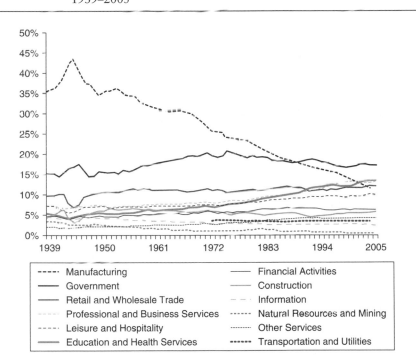

Source: Bureau of Labor Statistics

There are various explanations for this trend. Some argue that nearly all low-skill, low-wage manufacturing work is being funneled to "less developed" countries, whereas the advanced economy of the United States focuses on knowledge work and services (Bell 1973; Fröbel, Heinrichs, and Kreye 1982). However, it is also possible to argue that this simply reflects something "old"—the continued effort of employers to find the least expensive ways to produce goods (Cowie 2001). From this point of view, manufacturing remains central to the economy; however, it now takes place on a global scale, rather than on a national one. Yet another interpretation emphasizes that the United States is unusual—the decline of manufacturing employment is more pronounced here than elsewhere. Rather than reflecting a long-term, general trend away from manufacturing, the U.S. pattern may reflect a choice by American employers to seek low-wage sites for manufacturing rather than investing in improved techniques at home (Appelbaum and Batt 1992). All of these processes have played a role in shaping opportunity in the new economy.

Manufacturing is declining in the economy of the United States and other economically advanced countries, but should we conclude that we are truly post-industrial? In the new economy, manufacturing enterprises continue to employ more than 14 million American workers, approximately the same number of jobs that were present in this sector in 1970. Although manufacturing employs a smaller percentage of Americans than it once did, it remains a major force in the economy and creates demand for the products and services generated in other parts of the economy (Glyn 2005). It is not at all clear that manufacturing employment is in an inevitable, long-term decline to the point where it will entirely disappear. Rather, it remains an important, but less dominant part of what is now a more diversified economy.

Finally, the decline of manufacturing jobs does not mean the economy is "*post*-industrial," as Bell argued it would be. The practices of the old industrial economy have been woven into the development of the new economy. The fact that manufacturing opportunities have stagnated in the United States does not mean that the *ways of working* that developed in the old economy are on a path to disappearing as a result. We illuminate some of this evidence, first by considering the extent to which mass production is disappearing in the new economy.

The End of Mass Production?

Interpretations of the old economy generally agree that the core of economic activity centered on the production of manufactured goods (automobiles, steel, chemicals, appliances, etc.) in large quantities for mass markets. The Ford Model T is the classic example of what the American manufacturing economy produced—an affordable and highly standardized car, mass-produced by American workers in a central factory location (Chandler 1990). Coordinating hundreds (and sometimes thousands) of workers at a single location meant that employers such as Henry Ford had to develop bureaucratic management systems, complete with rigid job definitions, rules of conduct, and productivity expectations (Edwards 1979).

The dominant managerial approach of the time was to follow the practices of scientific management, which encouraged managers to replace skilled workers with cheaper, more replaceable low-skill workers, while removing discretion from the shop floor and placing it in the hands of management (Braverman 1974). Mass production, assembly-line manufacturing embraced this philosophy. Workers labored at repetitive, simple tasks, at a pace set by management and the technologies and bureaucracies under their control (Hounshell 1984; Meyer 1981; Noble 1984; Sabel and Zeitlin 1985). The result was the creation of legions of deskilled jobs, the dissolution of many craft skills, and a decline in the worker's ability to control the conditions and rewards of work (Braverman 1974; Pietrykowski 1999). This approach was enormously successful and formed the basis for the growth of the giant American manufacturing enterprises (Ford, General Motors, U.S. Steel, etc.) that dominated the American economic landscape and symbolized American economic power worldwide. This approach also fostered distrust and hostility between workers and their bosses, who developed an "us versus them" mentality, in which each side saw the other as having interests fundamentally opposed to its own. Thus, as managers tried various tricks to speed up work, those laboring on the front lines developed alternate approaches to try to restrict production (Burawoy 1979; Edwards 1979; Montgomery 1979; Tulin 1984).

If manufacturing is in decline, is mass production? To address this issue, consider work as it is performed in different "mega-sectors"—broad groupings of different types of economic activity. Each of these sectors make distinct contributions to the economy— in extracting resources, in processing resources, in delivering goods, and in providing services.[2] The trends for employment in these sectors are represented in Exhibit 2.2, which shows the growing importance of service sector work, as well as the proportions of the labor force employed in other industries that seem (on the surface) to have helped society progress beyond mass production. However, consideration of many of the jobs within each of these sectors highlights how the typical strategies for organizing work in the manufacturing-based economy have been exported to other sectors and shape how work is performed outside manufacturing.

For most of human existence, most workers engaged in the extraction of raw materials—working in the areas of farming, fishing, forestry, or mining. But by the early 20th century, these workers composed only a relatively small segment of the workforce. Those few who remain on farms today perform work that bears little resemblance to the pastoral ideals of the family farm. Rather, most farming occurs as part of agribusiness, in which the methods of mass production have been applied to the raising of livestock, poultry, and produce (Schlosser 2005). The extension of mass production

Exhibit 2.2   Employment Trends in Mega-Sectors: United States 1939–2005

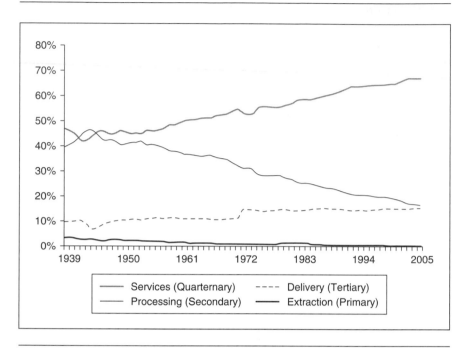

Source: Bureau of Labor Statistics

into farm work has required some farmers to learn to use advanced technologies to manage production. However, it also has contributed to the creation of a divided opportunity structure that limits prospects for workers (such as migrant farm laborers) to grow and advance. Mining remains an intensely physical activity, but again, it uses single-purpose machinery (such as "continuous miners") to grind the earth and place ore on conveyors—effectively, these are assembly lines that run in reverse.

The processing mega-sector focuses on the refinement of raw materials into finished goods, the intent of manufacturing and construction enterprises. As Exhibit 2.2 shows, processing work has declined significantly in the United States, which seems to support the "end of mass production" thesis. Yet, if we look at work within these sectors, mass production techniques have not been eliminated. It is true that there have been significant changes in this sector; for example, the traditional giant steel mill of the past has given way to smaller, more flexible "mini-mills" that produce smaller runs of more specialized steel products. But contemporary factories are still mass-producing consumer goods. And in certain parts of this sector, most notably the building trades, modern practices have made production *more* reliant on the use of standardized materials. Many homes are "prefabricated" in construction factories and simply assembled on site. Even the construction of "custom" homes (the largest of which are pejoratively termed "McMansions") depends heavily on cookie-cutter approaches to design and assembly. In short, flexible and custom production processes have taken hold in some areas, but they are being introduced in concert with a continued reliance on older systems of production, and are not simply replacing them (Pietrykowski 1999).

Even the delivery mega-sector (which includes transportation, wholesale trade, retail trade, and utilities) relies heavily on mass production techniques. For example, the United Parcel Service (UPS) and Wal-Mart (the world's largest private employer) use vast conveyor belt systems to sort and funnel parcels for delivery. Their successes are not built on unique products or customized services; rather, they are based on the application of mass production distribution techniques built with high technology and advanced accounting systems. Package sorting and shelf stocking are not skilled or challenging tasks that lead to career advancement; they are the type of routine work that one pursues as a means to earn a paycheck (McPhee 2005).

Finally, within the services mega-sector, which includes a wide array of enterprises—including leisure services, restaurants, hotels, real estate, financial, and public administration—mass production methods can also be found, often in highly developed, innovative forms. This sector includes some enterprises that do not rely on mass production—there are many relatively small enterprises in this part of the economy, and businesses such as high-end restaurants rely on workers' skills to produce a "unique" product for the consumer. However, this sector also contains many highly standardized operations that use the techniques of mass production to good effect. Each McDonald's, for instance, is little more than a small factory, composed of deskilled jobs designed to execute production of a limited array of standardized goods. Every aspect of the process, including the dispensing of condiments, the design of the stores, and the way in which customers are greeted, has been standardized so that the experience of eating (or working) in a McDonald's would be essentially the same wherever it was located. Even housecleaning teams are organized to work according to Taylorized methods, and the goals remain the same—to extract the maximum effort from each individual and minimize their chances of relaxing on the job (Ehrenreich 2001; Ritzer 1996).

In sum, there is a tendency to assume that the declining importance of manufacturing in the U.S. economy means that mass production is on the wane. It is not. Indeed, what has happened is that mass production techniques have been widely integrated into other sectors of the economy, including the rapidly growing service sector. Throughout the new economy, there has been an expanding reliance on advanced technologies to perform complex tasks. In some cases, these technologies have required an expansion in certain worker skills and transformed the ways in which work is performed. However, those same machines have also replaced many workers, both in manufacturing and elsewhere, and the persistence and spread of mass production methods cannot be ignored. The reality is that the new economy relies on *both* skilled work and deskilled work and uses both mass production techniques and newer, more flexible systems to produce goods and services.

New Cultures of Control?

Some observers of the new economy have argued that changes in technology, organization, and markets are transforming jobs, including manufacturing jobs, into more highly skilled jobs in small, high-tech, "flexibly specialized" enterprises (Piore and Sabel 1984). The workforce needed for this new kind of workplace will be highly educated, multitalented, and required to exercise creativity and decision making on the job. In the new economy, managers are cautioned against "micromanaging" and are advised to form workers into teams with their own team leaders. Even low-level employees are hired as "associates," implying that their input will be valued. Others are not convinced. They point to the experiences of low-level service sector workers, and agree with the critical philosopher Slavoj Žižek, who once quipped, "The employee of the month shows that one can be a winner and a loser at the same time." And, though acknowledging that new methods of organizing work are more common, these skeptics suggest that the reorganization of work is not just about introducing new markets or technologies, but also about developing sophisticated new methods to control workers and undermine their power in the workplace (Curry 1993; Parker 1985; Parker and Slaughter 1988). Thus, we must also ask, are new technologies and organizational designs increasing worker autonomy, creativity, and control?

Some of the most carefully conducted studies of this question have been performed by Steven Vallas, who has examined the introduction of computer technology and worker teams in the pulp and paper industry. His findings are important because they both support—and refute—core predictions regarding what happens to workers in organizations that use new technologies and managerial strategies. First, consider teamwork. Some analysts see worker teams as an ideological trick, a way of getting workers to believe that they have control when they do not. Vallas found that, rather than hoodwinking workers, work teams shared grievances, developed a heightened distrust of management, and developed class solidarities. And, by virtue of being a team, they expressed complaints to management with less fear of personal reprisal (Vallas 2003a).

But Vallas also found that the belief that teams produce shared decision-making responsibilities between workers and managers is exaggerated. In fact, the introduction of computer-regulating systems in the paper industry tended to increase the distinction between those who had the authority to make decisions and those who did not. Older workers interpreted the new technology as an affront to the craft skills they had developed through years of experience on the job, and expressed dismay at the new reliance on meters and printouts that provided information that they already possessed (Vallas and Beck 1996).

One of the important insights from Vallas' studies is that the new economy seems to be marked less by fundamental shifts in the amount of control workers have, and more by modifications to systems of accountability. In the old economy, the assumption was that workers should have no control and that jobs and machines should be rigidly designed and managed from above. In the new economy, where commitments to Total Quality Management foster a drive to work with exacting perfection, and where the responsibility for creating this outcome is placed on the shoulders of worker teams, a new dynamic of collective pressure is introduced. Workers now have increased responsibility for production, but not control over many of the decisions that shape it. These findings correspond with a number of other analyses that show that workplaces in the new economy do not operate on trust and cooperation and that workers and managers remain skeptical of each other's intentions and motivations (Appelbaum and Batt 1992; Osterman 2001; Parker 1985; Rinehart, Huxley, and Robertson 1997; Smith 1990).

Another supposedly new dynamic is the re-creation of craft communities, which some believe are reshaping the ways work is understood and performed. For example, the success of Silicon Valley enterprises has been attributed to their operating in a regional community of similar companies that specialize in computer work (Pietrykowski 1999). In some ways, these regional centers operate in a manner similar to the craft communities of bygone eras, and position workers to exert control through the development of a collective culture of standards and expectations. The work of skilled programmers in this environment is truly a team effort and requires high levels of skill and control, and workers are paid handsomely. On the surface, it looks like a model example of optimistic predictions about what work can become.

Questions have been raised, however, about how real these "craft communities" actually are. Even though many firms operating in Silicon Valley are small and nimble, they do not operate independently of larger corporate concerns (Harrison 1997). As we discuss further later in this book, some elements of life in Silicon Valley appear more enslaving than liberating. For example, programmers in Silicon Valley have developed a hyperindividualistic culture, where members of the work community believe it to be their moral responsibility to labor 12- to 14-hour days and sacrifice their lives for their jobs. They work long hours not simply because of managerial demands or because of coworker pressure, but because they have internalized a sense of commitment to the organizational culture (Shih 2004). Even when they can take time out—for example, to care for a newborn child—they often don't because it would undermine their ability to be "a player" in their work community (Hochschild 1997).

The End of Organized Labor?

Another possible indicator of an emerging new economy is shifting balances of power between workers and their employers. One means of assessing this dynamic is to consider trends in collective action and unionization. As Exhibit 2.3 shows, the rise of the old economy was accompanied by

Exhibit 2.3   Percentages of American Workers Who Were Union Members: United States 1930–2004

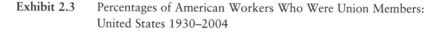

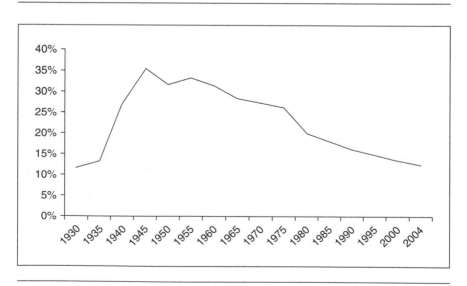

Source: Statistical Abstracts of the United States

a dramatic increase in union membership. At the middle of the 20th century, roughly 1 in 3 American workers belonged to a union. In the latter part of the century, however, membership plummeted, and by 2004, only 1 in 10 workers belonged to unions. The decline of unionization in the private sector has been particularly sharp; fewer than 8% of private sector workers belonged to unions in 2006. The one exception to the general pattern of decline has been in the public sector; government sector unions remain vigorous and have even grown, representing more than 36% of government employees in 2006 (Bureau of Labor Statistics 2007a).

Why this is the case remains something of a puzzle. One possible reason for falling union membership may be the declining proportions of the workforce employed in manufacturing, the sector in which unionization has traditionally been strongest. In addition, as American manufacturing employment has contracted, employers have become increasingly willing to close facilities or to relocate both within the United States and abroad. In some cases, this allows employers to move away from areas where unions exist to those where they do not; in other cases, efforts to unionize have been stymied by a company's threatened or actual relocation. However, nonmanufacturing workers do join unions in many other countries (including Canada). And the decline of unionization in the United States has occurred not just in sectors exposed to capital mobility, but also in sectors such as transportation and construction that cannot relocate. So, the real reason for declining unionization may be the failure of American unions to organize workers in the growth sectors of the new economy (Milkman 2006).

Much of the problem unions face in the new economy results from their limited success in attracting new members (Lichtenstein 2002; Moody 1997). American unions have been criticized for their lack of emphasis on organizing new groups in the post–World War II era, but even when they try, unions often meet with failure. One hindrance they face is that laws in the United States have made it hard to organize new unions. Although American workers won the right to organize with the New Deal–era Wagner Act (which also established the basic legal framework for collective bargaining in the United States), numerous restrictions have since been placed on unions and union organizing activities. The Taft-Hartley Act of 1947 was particularly important in this regard. This piece of legislation, strongly backed by an antilabor postwar U.S. Congress, eliminated some of the most effective weapons in unions' arsenals, most notably the closed shop (in which all workers at a particular place of employment are required to be union members), the sympathy strike (in which workers in one industry strike in support of workers in another), and the secondary boycott (in which unions attempt to persuade others not to do business with a particular firm whose workers are on strike). Even more importantly, Taft-Hartley made organizing new groups of workers much more difficult by authorizing states to pass "right-to-work" laws (which prohibit unions from making paying dues or fees a condition of employment), by strengthening employers' ability publicly to speak out against and resist unionization drives, by permitting strikebreakers to vote in union certification elections, and by making the process of union certification far less flexible than it had been.[3] In general, U.S. labor policy has not favored new union formation, which in turn has left many workers (especially those in the rapidly growing service sector) without the opportunity to bargain collectively (Fantasia and Voss 2004).

Moreover, employers have become increasingly hostile to unions and have successfully used a variety of tactics to discourage organization. In some cases, this involves the use of the "carrot," that is, making union membership less attractive by providing, voluntarily, some or all of the good labor conditions unions gain for their members (Milkman 2006). In others, it involves the "stick," that is, aggressive efforts to block unionization through methods such as firing organizers and the use of experts on keeping unions out (Goldfield 1987; Head 2004).

Some analysts have suggested that declines in union membership indicate that the new economy has the potential to change relationships between employers and employees for the better. New management philosophies and new forms of work organizations ostensibly promote collaboration and harmony between employers and their workers. Work in smaller enterprises involves direct face-to-face relationships between workers and their employers, which discourages a "them/us" mentality. Employers, increasingly dependent on workers' creativity and initiative, have learned to empower employees and to abandon the rigid managerial practices that encouraged mass unionization in the past. Encouragement is used to motivate workers, and the effort is to create company loyalties, rather than class loyalties (Kochan, Katz, and McKersie 1986; Piore and Safford 2006). In this culture of cooperation, traditional unions are not attractive to workers, who often see them as threatening the company's interests and their jobs (Heckscher 1988). Workers also are less likely to devote their entire careers to single employers, and instead build portfolios and move from employer to employer. In this context, unions representing particular workplaces become less relevant (Capelli 1997). But this is only part of the story.

The decline of unionization has occurred simultaneously with declining wages and job security. Thus, it may not be that changing management practices are making unions unnecessary; rather, it may be that unions have lost much of their ability to negotiate for workers' interests as their approach to organizing workers and challenging working conditions has been undermined by new organizational practices and changing economic conditions (Lash and Urry 1987). For example, American unions' ability to use the strike to put pressure on employers to raise wages or improve work conditions has largely evaporated. Consider that in 1970, 2,468,000 workers participated in 381 mass walkouts. In contrast, in 2005, there were only 17 large-scale work stoppages, and these only involved 171,000 workers (*Statistical Abstracts of the United States*).

The downward trend in union membership has also been witnessed in most countries outside the United States, but the extent of these declines has been variable. Scandinavian countries have retained remarkably high union membership. Countries such as Germany, England, and Australia have significantly higher union membership rates than does the United States (Exhibit 2.4). The United States stands out in the international arena because of the dramatic decline in membership and its exceptionally low current

Exhibit 2.4   Trade Union Members as Percentages of All Employees: International Comparisons

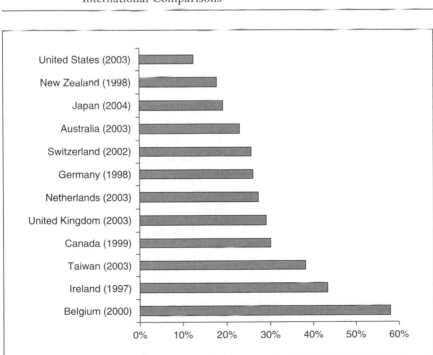

Source: Data provided to authors from the International Labor Organization.

levels. These declines have contributed to overall declines in wages, benefits, and job security for workers (Lichtenstein 2002). The fact that unions remain important in other advanced economies, and that service workers who are typically unorganized here belong to unions there, is strong evidence that unions can still matter. From this point of view, the conclusion is not that unions are a vestige of the old economy but rather suggests the need to consider how to reinvigorate the American union movement (Clawson 2003; Fantasia and Voss 2004; Heckscher 1988; Milkman and Voss 2004).

A New Global Economy?

A final and much-discussed aspect of the new economy concerns globalization. Has the emergence of a global economy fundamentally changed the U.S. economy and the situation of workers worldwide? In some respects, the global economy is not really new. The histories of virtually all modern societies, from the 16th century onward, can be traced to international economic ties (Wallerstein 1979, 1983). For example, colonial America participated in international trade of slave labor, sugar, rum, tobacco, cod, and textiles (Kurlansky 1998). Had it not been for these exchanges, the present-day demographic makeup of the United States would be profoundly different, as would its culture (Breen 1985; Sobel 1987). Likewise, the export of slaves to the United States had an enduring impact on the development of African societies (Fredrickson 1981; Portes and Walton 1981). Trade with Asia is not new, and the European "discovery" of America was the result of attempts to find better trade routes. Indeed, international trade has long been in existence and has gone through numerous cycles of growth and decline (Chase-Dunn, Kawano, and Brewer 2000).

Nevertheless, one must still acknowledge that the extent of global economic activity is unprecedented and that the penetration of global capitalism to all corners of the world is both more complete and more complex than ever before. The new global economy can be described as a vast international network capable of rapidly developing and diffusing resources, technology, and information across the world. Among the key characteristics of the new global economy are the following:

- The immense volume of trade and consumption between societies
- The rapid transmission of information between societies
- Powerful and transportable technologies implemented throughout the world
- Intense "dis-integrated" production, spread over national boundaries
- Flexible arrangements that enable employers to shift production and consumption from one society to another

The result has been cultural and technological changes in developed and developing countries accelerating at unprecedented rates. The speed of change, the extent of diffusion, and the flexibility of webs of connection set the current organization of work apart from the systems that preceded it (Castells 2000; Mattsson 2003; Milberg 2004). As one observer has remarked, the world is now, probably for the first time, approaching "universal capitalism" (Wood 2003).

Companies are integrating themselves into the global economy for a variety of reasons. Labor cost savings are a major motivator, but other considerations are also involved. Developing countries almost universally have lax environmental regulations, and this reduces the expense entailed in limiting and controlling pollutants (Jones 2005). The tax structures of nation-state systems also encourage the movement of jobs. For instance, when the industrial tool company Ingersoll-Rand relocated its headquarters to Bermuda, it reduced an annual $40 million tax bill owed to the United States government to $27,653 (which it paid to its new home country) (Johnston 2004). Companies also internationalize operations to secure government contracts and extend the global reach of their product lines.

The emergence of a global economy has had significant effects on workers in the United States and other industrialized countries, partly because the "national" character of companies has weakened and become ambiguous. American companies such as Ford have long had overseas operations in Europe. However, the scale of those overseas operations is something quite new. Ford (like nearly all the major employers in the new economy) is a multinational corporation that operates both within and beyond the political realms of nation states, and includes brands that used to be its foreign competition (including Jaguar, Volvo, Mazda, and Land Rover). Today, Ford employs more workers outside the United States than it does within, and it manufactures and sells cars around the world. Companies such as Ford can easily shift production from place to place, as workers in Michigan have discovered. When companies move or open new facilities abroad, they bring more than jobs—they also spread culture and methods of organizing work. A good example of this can be seen in the case of Japanese companies, which have built facilities in both the industrialized and developing worlds. These plants become vectors through which Japanese production methods diffused to places such as Marysville, Ohio, Northern Mexico, or Spain (Elger and Smith 1994).

The emergence of a global economy has undoubtedly brought with it much that is new. Workers in the United States are much more likely to encounter technologies and managerial practices that originated elsewhere, and workers in developing countries have been drawn into much more direct relationships with global webs of production. Still, arguing that all this is *entirely* new seems an exaggeration. Employers have relocated in the past, and even the earliest industrial firms in America "borrowed" practices from the pioneering British. It seems more accurate to say that globalization has accelerated and intensified existing dynamics at work. At the same time, the socioeconomic differences between the developed and developing world have not been erased by these changes, a theme we develop at greater length later. Nor have national differences in workplace practices been eliminated by globalization (Smith and Meiksins 1995). As with the various other changes we have reviewed, there is much of the old within the new economy and workplace.

The Old in the New: A Summary

Is there a new economy? The answer is both yes and no. The U.S. economy has shifted away from a near-exclusive reliance on its manufacturing base, and the expanding service sector is creating new job demands, as well as opportunities. Some new jobs require different skill sets than those needed in the industrial economy, but many do not. Reconfigured organizational designs, managerial philosophies, and technologies have expanded the need for skilled workers, and the workforce is more educated than ever. Work opportunities in the United States also are linked to the opportunities developing in other parts of the world. But the old economy is also embedded in the new economy, both in America and abroad. Charlie Chaplin's character from *Modern Times* could easily be transplanted into a present-day American McDonald's, Jiffy Lube, Wal-Mart, or UPS, as well as into the toy factories of China or call centers in India.

Class Chasms in the New Economy

As the practices of the old economy became integrated into the development of the new economy, did this reshape patterns of socioeconomic inequality? To answer this question, we focus on trends in earnings and the disparities in rewards between those workers laboring at the bottom and those highest in the class structure. We also consider the equally important question of mobility between classes, and the types of career ladders available to workers laboring on the front lines. Finally, we extend the discussion of inequality and mobility into a global and comparative framework, considering how global production is influencing life chances in developing countries and the disparities in opportunity provided by work in America and work abroad.

Class and Opportunity in the United States

Are the changes that we outlined earlier fundamentally reworking the American class structure in the latter part of the 20th century? From a class perspective, the presence of a new economy could foster new avenues for prosperity or create new paths toward poverty. One way of examining its impact on life chances is to examine trends in family incomes and how they relate to the emergence of a new economy.

Exhibit 2.5 offers a graphic illustration of the changes in income distribution in the United States. To make this graph, the families in the United States were divided into five equal-sized portions (termed "quintiles"). The shaded band at the bottom of the graph shows that in 2003, the poorest 20% of families subsisted on incomes ranging from $0 to $24,117, the next fifth had incomes from $24,118 to $42,057, and so on. Because the range of incomes for the top earners extends into the millions of dollars (for people like Bill Gates), we capped the graph at a maximum income of $180,000,

Exhibit 2.5   Incomes Received by Each Fifth, and Top 5% of Families: United States 1947–2003 (Converted to 2003 Dollar Values)

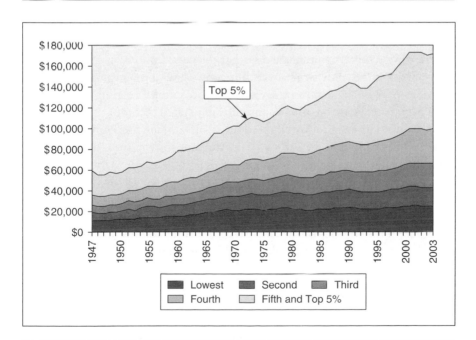

Source: Statistical Abstracts of the United States

but added a line representing the minimum income of the top 5% (around $165,000) in 2003.

One way of looking at this graph is simply to consider the relative positioning of each fifth of the population in 2003. Exhibit 2.5 shows us that roughly one in five families were living at or near the poverty level. The next fifth consisted of families that had incomes that offered scant possibilities for accruing savings and for whom homeownership was largely out of reach. But for those fortunate to be at the very top of the class structure, the transition to the new economy has presented golden opportunities to increase riches. According to the Bureau of Labor Statistics in 2003, the richest quintile earned *half* (48%) of the aggregate income earned by all families in the United States, whereas the poorest fifth of American families earned only 4% of the collective earnings (*Statistical Abstracts of the United States*). And these disparities have been growing during the 20th century. The AFL-CIO reports that in 1980 the average CEO made 42 times the pay of the average employee, but in 2005, CEOs were earning 411 times that amount (AFL-CIO 2007). It takes the typical American Wal-Mart employee an entire year to earn what the company's CEO makes before his morning coffee break.[4] These class divides, those separating the incredibly affluent from the rest of American society, have long been in existence and reflect the concentration of wealth and power that already existed in the old economy (Mills 1978). However, the chasms that separate those who have from those who do not are widening (not decreasing) in the American new economy.

One can present an optimistic case that the new economy is creating prosperity. Exhibit 2.5 shows that overall incomes have been rising throughout the latter part of the 20th century, but a more careful reading reveals that these overall gains are largely produced by expanded incomes among the top earners. In contrast, the lower three-fifths of families are earning roughly what families earned in the early 1970s. And, not shown in the graph (but as we discuss later), attaining these family incomes now requires the presence of two earners in many households, whereas in the early 1970s, single-earner households were more common. The conclusion is unmistakable—if the new economy is producing prosperity, it is doing so in a lopsided manner.

Although the top tier of the class structure is far removed from everyone else, an opportunity chasm divides members in the middle strata of workers from those at the bottom. Economists have termed this divide the **dual labor market.** The favored jobs are in the **primary labor market,** which offers higher pay, predictable career structures, opportunities for skill development (albeit limited for manufacturing workers in production jobs), and enhanced job security. In contrast, other workers are trapped in a **secondary labor market** comprised of jobs that offer low pay, unstable opportunities, and low security and that require little skill (Piore 1977). What is important to note is that this divided opportunity structure is not new; it developed within the old economy and has been transferred into the new one. Most low-skill jobs that workers hold today do not offer stepping-stones to greater opportunity or access to rungs on career ladders. These are dead-end positions that lack benefits such as health insurance, provide little security, and offer scant economic reward (Kalleberg, Reskin, and Hudson 2000). And in contrast to work in the old economy, far fewer workers laboring in these dead-end positions have the protections or pay offered through collective bargaining agreements that existed when more workers were union members.

What is more, in the new economy, some of the division between the primary and secondary labor markets is blurring—not so much because bad jobs are becoming good jobs but rather through the extension of insecurity into some jobs in the primary labor market. As we discuss later in this book, most workers—even those in good jobs—are working longer hours, facing new forms of insecurity, and experiencing greater challenges in locating work than was the case a few decades ago. Many of the good jobs are less good, and a variety of cultural and structural divides persist in limiting the prospects that people from lower-class backgrounds, ethnic minorities, or women, will obtain—and keep—these jobs (Perrucci and Wysong 2002).

Although the dual labor market is not new, what has changed is the extent to which career success in the new economy hinges on the attainment of a college or graduate degree (Levy 1998; Marshall and Tucker 1992; Reich 2002). As a result, Americans are entering college programs and returning to school at unprecedented rates. Exhibit 2.6 shows that a half century ago, fewer than 1 in 2 Americans earned high school degrees; today this figure has risen to nearly 9 in 10. College graduation rates have tripled during the same period, as younger Americans are increasingly likely to seek postsecondary education. In 2007, fully 57% of those aged 25 to 29 had at least some college education. And, as Exhibit 2.6 shows, in 2004, 1 in 4 (28%) Americans older than 25 possessed at least a bachelor's degree (U.S. Census Bureau 2004).

The dramatic increase in entry into college reflects personal responses to changing opportunity structures. As Exhibit 2.7 shows, in 2003, the average worker possessing a bachelor's degree earned twice the income of a high school graduate, and those possessing graduate level degrees earned considerably more than did those who ended their education at the bachelor's level. Those possessing only high school diplomas typically hovered at or near the poverty level, and the economic returns on their limited education have declined during the past three decades. After adjusting earnings to account for inflation, the typical male high school graduate at the end of the

Exhibit 2.6   Educational Attainment of Adults, Age 25 Years and Older: United
States 1960–2004

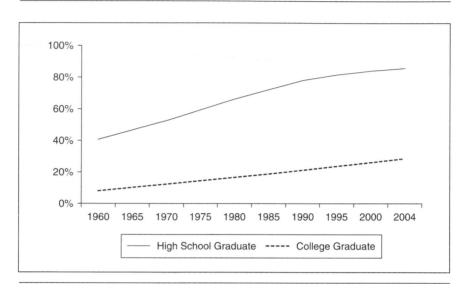

Source: Statistical Abstracts of the United States

Exhibit 2.7   Educational Attainment and Average Worker Earnings: United
States 2003

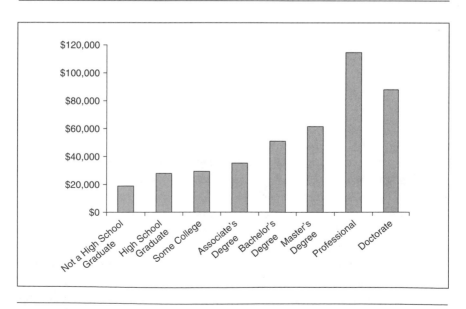

Source: Statistical Abstracts of the United States

20th century was earning considerably less than he would have in the early 1970s (Levy 1998).

The changing relationship between education and work is having a marked influence on the patterning of the life course. More and more members of the younger generation are delaying the start of careers as they pursue advanced degrees (although they also are likely to work at some kind of job while in school to pay for their education), and older Americans are returning to school in unprecedented numbers. More than one in three students enrolled in colleges and universities in 2003 were nontraditional ages, and most were women (*Chronicle of Higher Education Almanac* 2004). The most common reason for returning to school is to relaunch careers or adjust skills to keep up with today's job markets. Many workers find returning to school intellectually and socially invigorating, but it also creates considerable strains on marriages and family lives (Hostetler, Sweet, and Moen 2007; Sweet and Moen 2007).

Those who ignore the structural arrangements in the new economy argue that growing inequality is largely a result of people's choices—those who make the wise decision to attend college get good jobs, whereas those who do not fall behind (Herrnstein and Murray 1994). There are indications, however, that a college degree is no longer the guarantee it once was. The pursuit of advanced degrees may reflect a growing reliance on credentials for entry-level jobs but does not necessarily reflect the need for greater skills in the performance of these jobs (Brooks 2006). In other words, what used to be a high school graduate's job (e.g., administrative assistant) is now a college graduate's job. As the number of graduates has grown, more people find themselves underemployed, unable to find work commensurate with their education, and constrained in their abilities to move to where scarce jobs may exist. The growing supply of college graduates has created a buyer's market in which wages are stagnant or even in decline (Perrucci and Wysong 2002). The result is that, although college graduates outearn those with less education, not all groups of college graduates experience income gains or land good jobs (Bernhardt, Morris, Handcock, and Scott 2001). And ironically, as the value of education has increased, public resources directed to underprivileged students have been cut, leaving many individuals lacking the opportunity to attend college (Kahlenberg 2004).

It should be pointed out that the American pattern is in some ways unusual. American incomes, on average, have been stagnant in recent decades, whereas European average incomes have been gradually increasing. Similarly, the extreme emphasis on postsecondary education has not yet been replicated in Europe, although rates of university attendance are up and the demand for university graduates there is growing. However, what is common across all industrialized societies is the growth of economic inequality, indicating that the new class divides of the United States are not simply a matter of local peculiarities (Green 2006).

Class and Opportunity in the Developing World

The United States is not the only society that has been affected by the dynamics of the new economy. The global nature of production is reshaping employment and opportunity in developing countries as well. There is much disagreement about the ways in which the developing world is being affected, however. Some conclude that globalization is creating a global race to the bottom in which employers seek out those countries in which wages are lowest and worker protections are weakest, resulting in deteriorating work conditions and economic fortunes on a global scale (Cowie 2001; Lash and Urry 1987). Others disagree, arguing that globalization is expanding fortunes and creating opportunity, both at home and abroad (Friedman 2005).

However one assesses its effects, the global economy is transforming work in developing countries, influencing workers' opportunities and behaviors in the process. One such change can be observed in accelerated urbanization in developing nations. Exhibit 2.8 reveals that in 1950, for every one

Exhibit 2.8   World Population Projections (Millions): 1950–2030

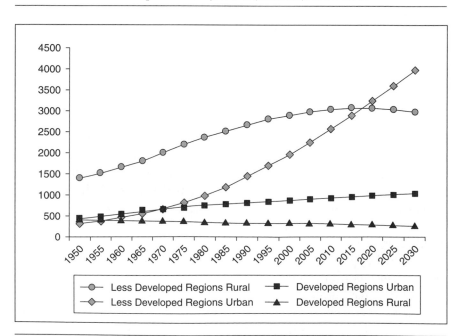

Source: United Nations, *Population Challenges and Development Goals* (2005)

person residing in a city in developing nations, there were three other people living in the countryside. By 2020, the developing world will have three times the population it did in 1950, and most of its people will be living in urban centers. In some respects, the causes of urbanization are similar to those that drove this process in the United States in the latter part of the 19th century. People are establishing communities that coincide with current opportunity structures. Urban centers are expanding because of push factors (the scarcity of jobs in rural areas) as well as pull factors (the perception that jobs will be available in the cities).

In some respects, this complex process of proletarianization (i.e., the transformation of peasants and other nonemployees into wage workers) parallels what happened in Europe and the United States in the 18th and 19th centuries (Deyo 1989). But in the developing world in the new economy, the problems of urbanization are magnified by scarcities of resources, by outmoded (and sometime nonexistent) infrastructures that are operating amid contemporary technologies, and by overpopulation. Some migrants, such as Chi-Ying (the young worker discussed in Chapter 1) successfully find factory work. But many cannot, resulting in the growth of huge concentrations of marginally employed people in places such as the *favelas* of Rio and Sao Paulo or the swollen urban centers of nonindustrial Africa. Grappling with the social problems associated with urbanization—including employment, health, safety, sanitation, and housing needs—is going to be one of the greatest challenges for the new global economy (Cohen 2003; Economy 2005; Kim and Gottdiener 2004; Roberts 2005; York, Rosa, and Dietz 2003).

Employment prospects in the urban centers of the developing world are often grim. There simply are not enough conventional jobs available to absorb the enormous mass of city dwellers. One result has been the growth of the "informal economy" in the developing world. This refers to a variety of both legal and illegal activities, including begging, drug dealing, making and selling food, various kinds of home work, operating jitneys and pedicabs, and many other jobs that are unregulated and unprotected. There is intense disagreement about the consequences of this type of work. Some see it as positive and as opening opportunities for urban dwellers to engage in entrepreneurship that can be translated to the formal economy. Awarding microloans to individuals, as practiced by Nobel Peace Prize recipient Muhammad Yunus' Grameen Bank, reflects this way of thinking about cultivating work opportunity in developing societies. Others respond that the informal economy actually competes with and eliminates more desirable employment opportunities. In this manner, it is "stealth" formal economy work—as when home workers produce goods that directly or indirectly make possible the production of goods that eventually are sold in retail outlets such as Wal-Mart and Costco (Davis 2006).

The movement of population within the developing world and the limited employment opportunities for migrants add to the pressure to move across national borders (Mattingly 1999; Sassen and Smith 1992). But even though jobs are highly mobile in the new economy, significant barriers discourage workers from moving across borders. The proportion of the world's population living outside its country of birth has only risen from 2% to 3% in the last 30 years, and, only about 10% of the population of the United States and Europe is foreign-born (Glyn 2005). Still, some migrants do come. For some, migration is a route to economic prosperity, as the experience of many immigrants in the United States demonstrates. However, there are also dangers. Once immigrants are removed from their local residences and social networks—and especially if they migrate illegally—they labor with fewer protections. This problem is dramatically illustrated in the lives of domestic workers from Indonesia, who serve as domestic workers in the Middle East. These young women are socially isolated, labor for meager compensation, and receive little protection from government agents in their home or host countries. In the worst cases, they are treated more like slaves than employees (Rudnyckyj 2004).

Are International Economic Divides Widening or Narrowing?

Decentralized production and internationalization have fostered the development of what is now a hallmark of the 21st century work—**global supply chains.** In the new economy, workers seldom create commodities, start to finish, at a central location within a nation state. More often, production is network-driven, divided up between companies that are spread across the world (Schrank 2004). Some of the most recognized names in the new economy—including Liz Claiborne, the Gap, and Nike—employ *no* production workers. Instead, these companies develop designs and marketing strategies and then contract work with independent manufacturers operating abroad (Gereffi 1994; Tracy 1999). In this system, Nike employees can be well paid and live in good stable communities, but the workers who produce Nike shoes often are not. The operations of these supply chains expand wealth in a progressive fashion, a dynamic illustrated in Exhibit 2.9. Relative to the profits attained at the final stages of production and distribution, workers laboring upstream (concentrated in the developing world) produce and accumulate smaller shares of wealth (Bair and Gereffi 2003). This dynamic leads to the interesting observation that the global economy may be simultaneously expanding wealth in both the

Exhibit 2.9   How Value Accumulates in a Global Supply Chain

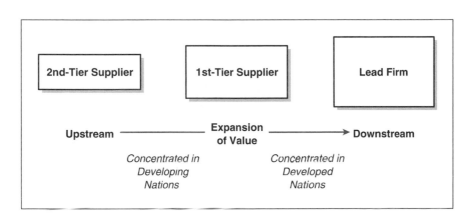

developed and developing worlds, but also expanding economic disparities of various kinds.

A few illustrations will help illustrate how multinational organizational structures and global supply chains affect international inequalities. We start with China, an emerging force in the global economy. How powerful is its role? The answer is commonly assessed by comparing **gross domestic product** (GDP), a figure (translatable to dollars) that indicates the total value of all goods and services produced within a country. With an annual GDP of $8,158 billion, China now trails only the United States and the European Union in its economic weight in the world system. In 2005, approximately 1 in 7 dollars (14%) of global economic wealth was created in China (Central Intelligence Agency 2005).[5]

The fact that China is catching up to the United States, Europe, and Japan as an economic power does not mean that its workers are in a similar economic situation to American workers. This fact is revealed if we consider **per capita productivity** in the United States and China. In 2003, China's contribution to the global economy was $1,100 per worker, compared with $37,870 contributed by the average American worker (*Statistical Abstracts of the United States* 2006). In part, these differences reflect the fact that workers in America and those in China are often engaged in different types of work, but it also reflects remarkable disparities in the wages received by workers who perform the same types of work in different regions in the global economy. For example, Ohio Art's famous Etch A Sketch used to be manufactured by workers in Bryant, Ohio, who earned $9 per hour before their jobs were outsourced to the Kim Ki

Corporation. This toy, along with most other toys available to American consumers, is now manufactured in China by workers who earn 24 cents an hour and who are expected to work 12 hours a day, 7 days a week. These workers receive no overtime pay (even though it is required by Chinese law), have no "weekends," and are docked wages whenever they miss work (Kahn 2003). Overall standards of living in China have risen in recent years, reflecting the fact that at least some Chinese workers benefit from the new economy. However, many Chinese workers are not allowed to migrate permanently from the countryside to better-paid factory jobs. They thus become migrant workers, temporarily housed in "dormitories," and eventually forced (often unwillingly) to return to their villages after a few years of labor (Ngai 2005). Levels of inequality *within* China have grown, however, indicating that there are both winners and losers in the process (Glyn 2005).

The economic gains that China experienced with its entry into the global economy have also come at a considerable environmental cost. As the land, water, and air quality in the United States became cleaner in the latter part of the 20th century (global warming effects excepted), the trend has been the reverse in China. Chinese rivers are now dangerously polluted, and five of China's seven major waterways are deemed unsuitable for human *contact*. In only 6 of China's 27 largest cities do residents have access to clean drinking water, and the wells of those living in rural environments have been fouled by industrial contaminants. The air is so polluted that airplane pilots frequently cannot see skylines when approaching Chinese cities (Economy 2005). The impact of China's industrialization goes beyond pollution. In its effort to harness waterpower for the industrial economy, the Chinese government developed numerous river dam projects that displaced millions of Chinese from their ancestral homes (Hessler 2003). Chinese economic development thus illustrates the complex effects of globalization— economic growth and rising living standards, combined with growing inequality, pollution, and community disruption. Chinese workers have reason to be both optimistic and concerned about their future.

Industrial development in places like China and Mexico has had particularly profound effects on the situation of women such as Chi-Ying. In many traditional cultures, women live in rigid patriarchal cultures where women's roles are highly circumscribed. Industrialization creates the potential for change because many of the workers who enter the factories of Northern Mexico and China are women. This experience could undermine patriarchy by altering women's roles and giving women a degree of economic independence. The reality turns out to be more complex. On one hand, women become more independent, and traditional forms of patriarchal domination by fathers and husbands have weakened in some ways. On the other hand, women in factories remain embedded in patriarchal kinship and community networks that restrict what they are allowed to do, and they encounter highly patriarchal work organizations in which women are the workers and men (often from their hometowns or part of their kinship network) are the supervisors (Lee 1998; Ngai 2005; Tiano 1994).

More than low-skilled manufacturing work is being directed to the developing world. India, for example, has experienced a remarkable growth in high-tech work during the last 20 years. Its successes can be attributed to a number of cultural factors, including strong work ethics and a large literate population fluent in English. The policies of the Indian government have also played an important role by directing resources to public schools, as well as to the training of engineers and medical specialists in state-supported universities. The International Institute of Information Technology (IIIT) in Hyderabad graduates 65,000 engineers and programmers a year (Bradsher 2002). As Thomas Friedman (2005) points out in *The World is Flat,* these workers are highly skilled and are eager to perform jobs that had been concentrated in America.

Jobs that use information and computer technologies in the developing world do not necessarily require workers to have technological sophistication. For example, one of the major new opportunities in India is call center work. Although call centers are relatively new to that society (emerging in the mid-1990s), their entry reflects a longer-term trend to disperse service work throughout the global economy. For example, in the 1970s, companies such as IBM outsourced consumer questions to call centers in Ireland and Scotland, as well as to lower-paying rural locales in the United States (Head 2003). By employing Indians, companies shave as much as 80% off the costs of performing this same work in America and Europe (Batt, Doellgast, Kwon, Nopany, Nopany, and da Costa 2005).

Call center employees tend to be young and well educated. Their work involves processing an unending stream of calls to and from American consumers. The shifts are long (8 to 10 hours/day, 6 days a week), and the pay is low by American standards ($50/week). And because of international time differences, most call center employees work at night. Although the jobs are clean, the work is repetitive and intense, involving rapid transitions between conversations that last only a few minutes. The job requires the skills to bridge cultural divides and the skills to navigate the protocols of response to customer queries, tasks that can take a month or longer to master. However, these skills are not transportable to other jobs beyond the call centers and do not offer strong prospects for upward career mobility (Batt et al. 2005; Taylor and Bain 2005).

Other high-tech jobs in India require far greater technological sophistication. An estimated half million skilled workers labor in the Indian software industry, making it the second largest pool of software engineers in the world (Khadria 2001; Mir, Mathew, and Mir 2000). The work of computer programmers is demanding, and they typically work 10-hour days (and often longer). On the surface, the daily tasks of computer programmers are similar in both India and in the United States. However, the *purpose* of their work is often different. Indians are commonly expected to write component pieces or debug systems within larger programs designed abroad. They are seldom engaged in the creative work that results in the development of a Google or YouTube that would directly benefit the Indian economy. As a result, India is unlikely to be positioned as a new knowledge creation center but rather, as an upstream link in knowledge creation, which will remain centered in America and other highly developed societies (Parthasarathy 2004).

The emergence of high-tech work is challenging traditional Indian customs and introducing Western cultural values. For example, traditional Indian norms largely prohibited young men and women from engaging in casual interaction, dating, and having romances in advance of arranged marriages. Call centers, with their comparatively high pay, mixed-gender workforce, and night work, are promoting sexual experimentation and increasing the number of "love marriages" among a young generation of workers. The long hours of work in the computer industries are creating the same types of stresses and overburdened schedules that American workers experience. In India, many of the strains of long schedules are buffered by the work of stay-at-home wives, as well as by grandparents who are willing and available to watch young children (McElhinney 2005; Patni 1999; Poster and Prasad 2005). But one wonders if the current generation of Indian professional women will challenge these intergenerational roles as their children have children.

Although the growth in the number of Indian programmers and engineers is remarkable, the popular image of high-tech India is somewhat misleading. In comparison with the United States, Indian high-tech workers compose an extremely small proportion of the overall workforce (there is approximately 1 skilled software engineer for every 2,200 Indians). Most Indian workers continue to engage in traditional agrarian work in rural villages or in menial jobs in urban centers. Day-to-day business involves contending with power outages and political corruption. Although work in India is strongly linked to the work being performed in Silicon Valley, the quality of life in India is still a far cry from the post-industrial experience of California (Chakravartty 2001; Patni 1999).

Conclusion

In this chapter, we tested a variety of assumptions concerning the emergence of a new economy. We concluded that there is indeed a new economy, but that this new economy has not abandoned the practices developed in the old economy. Instead, it has incorporated them into its operations. It has done so in complex ways, for example, by integrating some of the most oppressive aspects of industrial production job designs into service sector job designs. The new economy has pushed the long-standing practice of relocating jobs to places where labor is cheapest, but has done so in an accelerated fashion made possible by communication and information technologies. The desire to undermine workers' influence over how work is done and how it is compensated has remained intact, but the techniques to do so have shifted.

Although some have argued that the new economy is fostering a new era of opportunity at home and abroad, we are not convinced that this is the case. It is creating affluence, but in a manner that reshapes and reinforces opportunity chasms that separate those who have from those who do not. In American society, work in the new economy is shaped by two key divides. First, those at the top enjoy a radically better quality of life than does the rest of the society. The second divide separates those in the middle, who have jobs that are both better and increasingly fragile, from those at the bottom, whose prospects for career advancement are largely blocked by the design and allocation of opportunity. And the transformations that shape the opportunities at home are reshaping the opportunities abroad. Those fortunate enough to occupy good jobs in developing societies can enjoy a quality of life remarkably similar to that of the fragile middle in America. But once one considers the types of jobs being performed by those at the bottom in developing societies, and the environmental conditions in which this work occurs, the perception of a flat world disappears and the existence of chasms becomes all too apparent.

Are workers becoming more or less powerful? Are their work conditions improving or deteriorating? Is alienation on the increase or decline? Advocates of either perspective can gather numerous examples to support their beliefs. This is possible because economies have become far more diverse than they were in the old economy, not only in terms of what is expected of workers, but also the returns that work offers. Which workers are we talking about, and in what types of jobs? Considering these dynamics reveals that there are *multiple trajectories* at play in the new economy, trajectories influenced by social class, but also by gender, race, and nationality. In the remaining chapters of this book, we focus on the issue of work *diversities* and *divides*, both of jobs and workers, as well as the *tensions* that are emerging in the new economy.

Notes

1. The Bureau of Labor Statistics (2006b) reports that in 2005, 2,952,740 people were employed in computer and mathematical occupations in the United States, and 10,797,700 in food preparation and serving related occupations.

2. Economists use the terms "primary," "secondary," "tertiary," and "quaternary sectors" (respectively). However, because they also use the phrases "primary" and "secondary labor markets" to describe divided opportunity structures, we use the terms "extraction," "processing," "delivery," and "service provision" in accordance with the divisions identified by Kenessey (1987).

3. The United States employs a cumbersome, two-step process for union certification in which workers must first sign cards indicating their interest in forming a union, then, at a later date, vote. Under the Taft-Hartley Act, all of this must take place under the watchful bureaucratic eye of the National Labor Relations Board. As this book was being written, an attempt to loosen this process by creating a one-step election system was being proposed again in Congress.

4. Wal-Mart CEO H. Lee Scott raked in $15,681,507 in compensation in 2005 (AFL-CIO 2007).

5. The U.S. GDP in 2005 was $12,370 billion; the European Union GDP was $12,180 billion. The United States and the European Union each contributed 20% to the total world gross domestic product (Central Intelligence Agency 2005).

3

Gender Chasms in the New Economy

ileen's (the professional engineer discussed in Chapter 1) biography illustrates new patterns of gender inclusion in the new economy, as well as the challenges confronting women negotiating the demands of work and family. Her experiences reflect those of the baby boom generation—and now some of their daughters—who entered the American workforce in unprecedented numbers in the latter part of the 20th century. On one hand, Eileen's story is one of liberation because she successfully entered into a career that had been largely sex-segregated in the old economy. On the other hand, many aspects of Eileen's life seem anything but liberated because work takes its toll on her ability to find the time to spend with her children and to have a life outside of her job. But is Eileen a typical female worker in the new economy? Consider that Jamal's mother—a poor African American woman with drug dependency problems—matured with the same birth cohort as Eileen. Her different circumstances, and those of her son, illustrate how the gendered experiences of work in the new economy remain intertwined with the consequences of race and class.

This chapter links gender with work to consider how the opportunity structures in the new economy shape the careers of men and women. We examine the impact of women's entry into various occupations, as well as men's reluctance to perform "women's work." We argue that the new economy has expanded and diversified employment opportunities for both women and men, but also that enduring forces continue to segregate workers by gender and to impose hurdles that block mobility for many women workers. Some of the chasms that separate male and female workers can be attributed to the choices people make concerning what careers to pursue and where to concentrate their energy (in the home and in the workplace). But these gaps also are established by differential treatment on the job and by the various ways workers are subject to discrimination. Some inequalities are the product of interpersonal dynamics, but others result from job designs, as well as the value placed on jobs that were performed by women in the old economy. As the shapers of the new economy grapple with the need to support mothering—and fathering—in a society where most adults are expected to work outside the home, the multiple causes of gender inequality at work present critically important policy concerns. We address these issues, examining the gender-work connection from comparative perspectives that include class, racial, and international variations.

When Did Home Work Become Nonwork?

Our friend Erika is a stay-at-home mother who left her job to support her husband's career and raise their two young children. While Kevin is at work in his research lab at a prestigious university—a job that requires long hours and frequent travel—Erika manages the home front—tending to scraped knees, arranging playdates, and doing all the other things that keep their family integrated into their community. For his efforts, Kevin receives unemployment insurance, retirement benefits, and a handsome salary. Although she receives occasional pats on the back from her friends who speculate about "how difficult it must be to stay home with the kids all day," Erika's social status is decidedly lower than that of her husband. She receives far less social recognition for her efforts, no awards, and no pay. Her status as a nonworker is most tellingly revealed when she is asked about her intentions to "go back to work" after her children get older. Her numerous frustrations with mothering work (which are more serious than her culture admits) are felt alone and in private. As a result, Erika is beset with ambivalent feelings about her career choices. Though taking pride in her work as a mother, she misses the rewards of having a good job in the world of paid work. And her husband also loses out, not in the workplace, but in having only brief windows of time to spend with his children.

Erika and Kevin have what is commonly considered to be a "traditional" household arrangement. In reality, when viewed historically, this clear-cut separation of husbands' and wives' roles is anything but traditional. For most of human existence, almost all work centered in and around the home. In these **household economies,** there were gendered divisions of labor, with some tasks primarily assigned to women (cooking and child care being the most notable), but both genders contributed to the family economy. Shared responsibilities were also common. For example, in colonial America, when husbands became ill, wives served as "deputy husbands" and assumed responsibility for virtually all the activities previously performed by their spouses. Although women were not the political equals of men, they were considered workers and their efforts were considered absolutely essential to family survival (Boris & Lewis 2005; Boydston 1990; Ulrich 1982). Furthermore, the culture did not make a distinction between "going to work" and "going home," because work and family life were, to a great extent, one and the same. As a result, women who worked in and around the home were defined as being real workers, and their efforts were visible and socially recognized. A close approximation of agrarian work–family life worlds can still be witnessed on family farms, where the work of wives and husbands intertwines in proximate physical spaces (Cohen 1985).

Following the industrial revolution, and during the late 19th and early 20th centuries, American culture reconfigured its orientation toward home-work and embraced a new **ideology of separate spheres**. But this did not happen right away. In fact, when the early factories physically separated paid employment from work around the home, the first workers were often women and children. Men sometimes refused to participate in these new arrangements (Hareven and Langenbach 1978). However, within the two succeeding generations, men's and women's roles became sharply differentiated. The new system cast men into the role of wage earners, which in turn encouraged them to evaluate self-worth in terms of career success and the ability to provide for their family's economic needs. Many women, however, were expected to stay in the home and tend to the needs of their spouses and children. Although this husband/breadwinner–wife/homemaker arrangement is commonly termed "traditional," it is actually a modern arrangement unique to the new industrial order. Its influence had major effects on divisions of work in the home, women's access to jobs in the paid economy, and social policy. Women were cast as the weaker sex, were considered dependents rather than workers, and were thought of as being less capable of "real work" than men. A landmark Supreme Court decision (*Muller v. Oregon,* 1908), for example, ruled that it was not only acceptable, but also desirable, for employers to limit the number of hours their female employees worked. These are the words of Justice David Brewer, who wrote the majority opinion:

> That woman's physical structure and the performance of maternal function
> place her at a disadvantage in the struggle for subsistence is obvious . . . and, as

healthy mothers are essential to vigorous offspring, the physical well-being of a woman becomes the object of public interest and care in order to preserve the strength and rigor of the race. (Quoted in Boris and Lewis 2005, p. 81)

Justice Brewer's comments reflected the dominant cultural attitudes of the early 20th century—that a woman's place is in the home. This ideology of separate spheres not only assigned home work to women and legitimated dis-crimination against them, it also redefined the economic value of women's efforts in the home. Rather than the home being a place of work (as it was in the 18th century and before), the home came to be viewed as a haven from work—the place where men recovered from the toils of the factory and the office (Lasch 1995; Zaretsky 1986). What was previously considered work was recast by an emergent **cult of domesticity** that asserted that household tasks were something other than labor, could be effortlessly performed, and offered so many intrinsic rewards that financial compensation was not necessary.

Even today, some cultural and political leaders call for a renewal of pol-icy to support the supposedly natural arrangement of the husband/bread-winner–wife/homemaker model (e.g., Santorum 2006). Their vision of a traditional family corresponds with how families were represented on tele-vision shows that aired in the mid-20th century, such as *Leave it to Beaver, Ozzie and Harriet,* and *Father Knows Best.* These archetypes were of young, healthy, self-sufficient, heterosexual couples that could prosper on the efforts of one breadwinner. But as Betty Friedan documented in her path-breaking book *The Feminine Mystique* (1963), home work in the 1950s bore little resemblance to the image presented on television. For most stay-at-home mothers, home work entailed a grinding repetition of alienat-ing tasks, social isolation, and subservience. Idealizing these families, and calling for a return to the "traditional" family structure, reflects inaccurate ahistorical assumptions about the gendered division of labor and the orga-nization of household work.

Calls for a return to "tradition" also ignore the reality that, even as late as the 1950s, the adoption of the husband/breadwinner–wife/homemaker arrangement was only possible for two-parent families who had opportuni-ties to get good jobs. This arrangement was not an option for those who were working in low-paying jobs, which included many immigrants and persons of color (Coontz 1992, 1997; Gerstel and Sarkisian 2006). White middle-class men typically earned incomes that enabled them to support an unpaid wife, but working-class and poor men, particularly immigrants and racial minorities, did not. The notion that women should be "free" from work to tend to matters in the home was not applied to these groups, whose homes were not protected or idealized. African American women were clearly expected to work as servants in support of the new ideal household arrangements for a privileged white society (Dill 1988; Glenn 2002). Early 20th century immigrant women were also far more likely than were their native-born counterparts to work outside the home. When they attempted to conform to the dominant view that women should be homemakers, they encountered financial hardship (because their husbands' incomes were often low). In response, many found ways to make significant economic contributions to the household *within* the home by taking in laundry, managing boarders, or by producing clothing and other necessities (Amott and Matthaei 1996).

In sum, the problems faced by Erika are a legacy of her culture's adaptation to industrialization, but are also specific to her racial identity and class position in the new economy. For women in less desirable economic circumstances, the prospects of choosing to stay home and mold their lives to conform to the mythic "traditional" arrangements were (and are) not commonly available. Since the 1950s, expectations regarding middle-class white women's paid employment have changed significantly and the cult of domesticity has weakened. However, the assumption that women will have primary responsibility for the management of the home remains. Moreover, this work is still not recognized for the vital economic role it plays in preparing the next generation for work and enabling workers (like Kevin) to put in long, undistracted hours on the job (Crittenden 2001; Hochschild and Machung 1989; Kanter 1977).

Women's Participation in the Paid Labor Force in America

Although the husband/breadwinner–wife/homemaker arrangement was only available for a portion of the families in the United States in the old economy, it was a dominant model for organizing family lives. Women, especially married women with children, were far less likely to work than were men. And, although this arrangement presented very different opportunities, the clear-cut gendered division of labor created an effective means of maximizing collective family resources for the middle class (Becker 1981).

Today, this arrangement has become increasingly rare because women are almost as likely to work as men and most married-couple households contain two working adults. Exhibit 3.1 shows the magnitude of this change. In 1940, only one in four women was in the paid labor force, but by 2005, nearly two in three were. Men's participation declined slightly, reflecting the aging of the population (there are now more retirees) as well as delayed entries into the labor force.

Exhibit 3.1   Men's and Women's Labor Force Participation Rates
(Age 16 Years and Older): United States 1940–2005

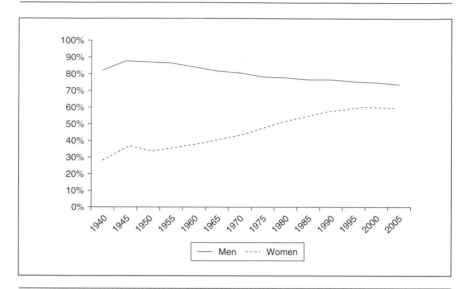

Source: Statistical Abstracts of the United States

Exhibit 3.2 shows that the normative arrangement of the old economy, in which the husbands went to work and the wives stayed home—is now the exception. In 2004, most married couples in the United States were dual earners, with both the husband and wife in the paid labor force. And this arrangement is common for families throughout most of the adult life course.

There are several reasons for the increase in women's labor force participation and the growing numbers of dual-earner couples. First is the changing perception of women's place in society and the remarkable transformation in gender role ideologies. But changes in the structure of the economy have been equally important. Stagnant male incomes made it increasingly difficult for the average husband to support a family on his wage alone (a situation long familiar to immigrant and minority families). Married women, whether feminist or not, found that they *had* to work if their families were to maintain their expected standard of living (Bernhardt, Morris, Handcock, and Scott 2001; Gerson 2001; Warren and Tyagi 2003). Also, the remarkable growth of the service sector (a significant employer of female workers even in the old economy) created a demand for an expanded female labor force in jobs that do not challenge conventional attitudes about gender (Reskin and Roos 1991).

Exhibit 3.2   Employment Configurations of Married Couples: United States 2004

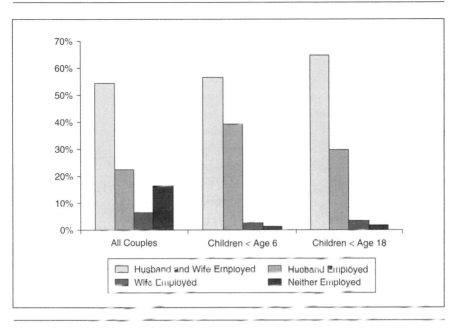

Source: Statistical Abstracts of the United States

The rise of the dual-earner couple has forced families to confront issues that were less commonly present in the old economy, such as who will care for children while both partners are at work, whose career will take priority, and how to select and pay for day care. As we discuss later, old gender templates play a strong role in determining couples' responses to the stresses they experience in the new economy.

Gender Inequalities in Compensation

Men and women still face very different prospects of ever rising to the top of career ladders, falling off those ladders, or even making it beyond the lowest rungs. The data are sobering. On virtually every measure of earnings, women trail well behind men. In 2004, for every dollar a man earned, women still only earned about 76 cents. The average woman who worked full time earned $9,575 less than the average full-time male worker. Women are twice as likely to be poor and are much more likely to hold low-paying jobs that offer slim prospects for upward mobility (Gilbert 2003; Kalleberg, Reskin, and Hudson 2000; Padavic and Reskin 2002). They are half as likely as men

to make it into the ranks of top income earners, and for every woman holding a position as chief executive officer (CEO) of a major corporation, there
are nine men (Bureau of Labor Statistics 2006a; Zweigenhaft and Domhoff
1998). Simply stated, female workers have not caught up to men in the new
economy, and recent trends do not indicate that the gaps are narrowing as
much as one might hope.

There are many complexities to analyzing the extent of gender inequalities in earnings. For example, should one compare all women with all men,
or only those who are working in the paid labor force? Note that the latter
analysis tends to underplay the extent of gender inequality because women
who are not in a position to earn any income are excluded from the calculations. The fact that women are much more likely to be working in part-
time jobs (often because of their scripted gender roles) creates additional
biases in wage disparity calculations that base analysis on full-time workers. But even when the comparisons of male and female workers are
restricted to those who are working full-time, year-round jobs, women still
tend to earn considerably less than their male counterparts, a dynamic
revealed in Exhibit 3.3.

Exhibit 3.3   Men's and Women's Earnings: Full-Time, Year-Round Workers:
United States 1960–2004

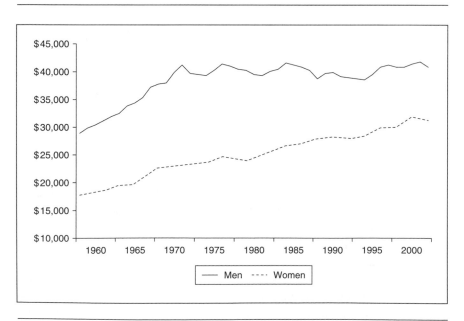

Source: DeNavas-Walt, Proctor, and Lee (2005). Adjusted to 2004 dollars.

These findings represent bad news, but there are some encouraging signs. Although women still lag behind men in earnings, the disparity is considerably smaller than it was in 1960, when the average full-time female worker only earned 60 cents for each dollar men earned. Unfortunately, some of women's gains have been the result of the stagnation in men's wages. Also, the most recent decade of data suggests the rate of convergence between men's and women's earnings is slowing, at least in comparison with the gains made from 1970 through 1990. The reasons for this slowing convergence are not altogether clear, but one study suggests that it may be the result of contemporary women pursuing jobs that offer lower pay, a dynamic we discuss later in this chapter (Blau and Kahn 2006).

In sum, there are wide gaps between men's and women's earnings. Some of the explanation can be found in the fact that household work is not compensated, which leaves many women without paychecks. The expectation that women will assume disproportionate household and child care responsibilities increases the odds that women will be funneled into part-time jobs, which also results in lower earnings. But even among full-time workers, women's pay lags well behind that of men. Why is this?

Socialization, Career Selection, and Career Paths

The large-scale entry of women into the paid labor force has reshaped how men and women view their roles and capabilities as well as opened doors that were previously closed. As a result, the gender composition of entire fields has shifted. For example, women are now the majority of graduates in veterinary colleges, a profession that in the 1960s and before was exclusively male. And men are now more apt to express a strong desire to spend time at home with children and are beginning to enter fields (such as nursing) where they were previously nearly entirely absent. Nevertheless, gender segregation remains a central part of the contours of work in the present-day American economy. One study estimated that half of all American working women would have to change fields to eliminate the disparities in the gender composition of occupations (Jacobs 1999).

One reason for these imbalances is that gender templates encourage men and women to have different aspirations. From early childhood, individuals make choices about what kinds of activities to engage in, what skills to develop, and what interests to pursue. Along each step of the way, they develop skills that, in turn, become part of the toolkit they use to construct future encounters. Ultimately, these choices influence eventual occupational goals and destinations. However, these choices are made in the context of powerful social expectations about gender. Various **agents of socialization** (including parents, schools, and the media) instill beliefs that boys/men and girls/women are not equally suited to all tasks. The confluence of social pressures and the transmission of taken-for-granted paths, over time, explains why men and women place different emphases on their identities as breadwinners, why they select different occupations, and whether they are willing to sacrifice their careers for the needs of their children and spouses. The result of socialization is that boys and girls (and men and women) are continuously molded to have interests and personalities suited to different types of endeavors (Bem 1993; Chodorow 1978; Lorber 1994; Lott and Maluso 1993).

Americans are socialized to expect women to fill certain positions in society, and in that respect many jobs are "gendered." Not every culture views these lines of work in the same way. For example, in the United States, most secondary school teachers are women, but in India, four in five secondary school teachers are men (United Nations Statistics Division 2005). Similarly, 71% of doctors in the United States are men, but nearly three in four Russian doctors are women. In other words, expectations about which jobs should be done by men or women are culturally, not biologically, determined (Harden 2001). Although women have made remarkable forays into many professions that were previously nearly exclusively male, Exhibit 3.4 shows that many occupations remain almost exclusively female. Notably, nearly all preschool and kindergarten teachers are women, as are secretaries, dental hygienists, dental assistants, dieticians, typists, and child care workers. Conversely, women are nearly entirely absent in construction industry jobs such as those performed by plumbers, carpenters, and electricians.

Gender expectations interact with class and racial differences to produce a complex, varied pattern of gender segregation in American workplaces. For women at the bottom of the opportunity divide, a major problem is the existence of **occupational ghettos,** gendered jobs that typically offer few paths to upward mobility (Grusky and Charles 2004). As Exhibit 3.4 shows, most maids, day care workers, and secretaries are women, and women from virtually all racial groups tend to be clustered in low-paying occupations such as these that offer few avenues for upward mobility. Indeed, the fact that the income gap between white and black women is relatively small (far smaller than it is for white and black men) is linked to the fact that women of all races are funneled into lower-paying jobs.

Socialization encourages workers to feel well matched to gendered lines of work. For example, one study of secretaries found that most of these women (especially older women) enjoyed their jobs. Rather than perceiving gender ghettoization as a source of oppression, they had accepted what they

Exhibit 3.4   Occupations With the Highest Percentages of Female Workers:
United States 2005

Occupation	Percent Female	Number Employed (1000s)
Preschool and kindergarten teachers	98%	719
Secretaries and administrative assistants	97%	3,499
Dental hygienists	97%	132
Dental assistants	96%	259
Dietitians and nutritionists	95%	68
Word processors and typists	95%	295
Child care workers	95%	1,329
Licensed practical and licensed vocational nurses	93%	510
Occupational therapists	93%	85
Receptionists and information clerks	92%	1,376
Registered nurses	92%	2,416
Speech-language pathologists	92%	98
Hairdressers, hairstylists, and cosmetologists	92%	738
Payroll and timekeeping clerks	91%	164
Bookkeeping, accounting, and auditing clerks	91%	1,456
Teacher assistants	91%	947
Maids and housekeeping cleaners	90%	1,382
Human resources assistants	89%	66
Health care support occupations	89%	3,092
Billing and posting clerks and machine operators	89%	427
Nursing, psychiatric, and home health aides	89%	1,900
Tellers	87%	418

Source: Statistical Abstracts of the United States

were socialized to believe—that men and women are different from one another, that women are skilled at doing tasks that men are bad at performing (especially nurturing), and that they themselves possessed these skills (Kennelly 2002, 2006).

Men with lower levels of education traditionally enjoyed advantages over their female counterparts in the old economy. These men were encouraged to seek, and generally found, jobs in manufacturing, construction, or automobile repair—work that offered opportunities for skill and income expansion and stronger prospects for economic security in both the short and long term. But less educated African American men were a notable exception because they often were excluded from jobs of this kind.

In recent years, the perception has grown that the situation of boys from poor and working-class backgrounds has deteriorated and that girls now have the advantage. The decline of well-paid manufacturing jobs, and evidence that girls are outperforming boys in school, seem to indicate that boys who do not have high levels of education are aspiring to jobs that no longer exist. However, a careful look at the evidence indicates that race, rather than class, is interacting with gender aspirations here. Less educated white men are not aspiring in large numbers to enter traditionally female fields but continue to find desirable, "masculine" work in areas such as construction or criminal justice. Poorly educated African American men (like Jamal, the young worker introduced in Chapter 1) continue to hope for traditionally male jobs that are unavailable to them and thus are falling behind (Kimmel 2006; Young 2003). Gender segregation at the bottom of the opportunity divide persists, in other words, but takes different forms for members of different racial groups.

Gender socialization also plays a role for men and women holding comparatively good jobs in the new economy. One means of illustrating its power is to consider how young adults select college majors, a critical step in determining subsequent career options. Exhibit 3.5 reveals that even today, young men continue to gravitate to traditionally male-dominated fields (e.g., computer science, engineering) and women tend to gravitate to college majors associated with helping professions (psychology, education, health care). In part, this is the result of gender beliefs, one of which is that men tend to be better at math than women. Research into this issue reveals that differences in mathematical abilities are actually small, or even nonexistent (Hyde, Fennema, and Lamon 1990). But these beliefs create self-fulfilling prophecies. Once the perception of capabilities is embraced by young men and women, even when those individuals have equivalent skills, it influences their aspirations to pursue mathematics-related professions. In other words, boys do not pursue math-related careers because they actually have better skills

Exhibit 3.5   Gender Compositions of Bachelor's Degrees Conferred: United
States 2002

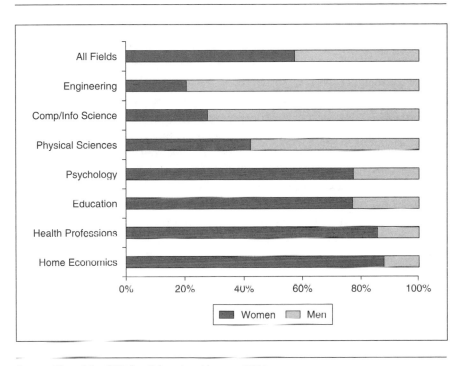

Source: Chronicle of Higher Education Almanac 2004

than girls; they do so because they *believe* that they are better (Correll 2001, 2004). But this is not the only factor in play. As we discuss later, those women who do enter male-dominated fields (such as engineering and science) are commonly marginalized, or find that these jobs are incompatible with their family responsibilities (Committee on Maximizing the Potential of Women in Academic Science and Engineering 2006; Preston 2004).

Aside from its role in career selection, gender socialization influences how men and women respond to work and family strains. Among working-class and poor families, women are more likely than men to arrange work schedules to allow involvement in their children's lives (Garey 1999). Among higher-income families, men seldom sacrifice their careers for family (Becker and Moen 1999). To compensate for their husbands' unwillingness to alter their career goals, women modify their work lives, scale back their work hours, and redirect their professional interests toward alternate careers (Moen and Sweet 2003; Wharton 2002). As a result, dual-earner families are transforming the 20th century separate spheres model into new **neo-traditional arrangements,** wherein women retain primary responsibility for child care while keeping one foot in the labor market in lower-pressure (and typically lower-paying) jobs.

The strategic choice to scale back work hours comes with significant costs to professional women, not only in immediate compensation, but also in long-term career prospects. Working women who are able to arrange shorter hours or flexible schedules with their employers are often placed on **mommy tracks** and are assigned tasks that offer fewer rewards and less opportunity for growth compared with workers on the fast track (Barnett and Gareis 2000). And even when more attractive opportunities to scale back are present, which can be the case for professional workers such as lawyers and academics, there is a very real prospect that they will never be able to get back on track once they step off (Meiksins and Whalley 2002; Moen and Roehling 2005).

For dual-earner couples, neo-traditional arrangements create marked disparities between wives' career opportunities and those afforded their husbands. As one partner's job is put on hold, the other's continues to grow, which reverberates into subsequent life course decisions, such as whose job to favor, when and where to move, and who should assume responsibility for family care work. Exhibit 3.6 illustrates this dynamic. When dual-earner couples manage their relationships, they can choose to favor either

Exhibit 3.6   Percentages of Husbands and Wives Reporting That Their Career Was Favored Over Their Spouse's Career (by Life Stage): Dual-Earner Professional Couples

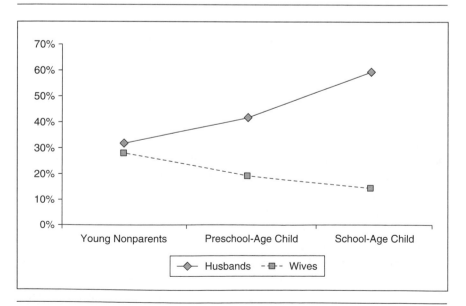

Source: Sweet and Moen (2006)

one partner's career (leaving the other as a trailing spouse), to take turns, or to give priority to neither partner's career. Most husbands and wives start their careers on relatively equal footing, with neither partner's career being favored, but as their lives develop over time, wives' careers tend to be towed along and follow the direction charted by their husbands' job opportunities. Their options are constrained by investments in husbands' careers, which in turn limits their options to advance and grow, even after children have left the household (Bielby 1992; Bielby and Bielby 1992; Pixley and Moen 2003; Sweet, Moen, and Meiksins 2007).

Catherine Hakim (2001) argued that the reason for this type of dynamic is that many women express a preference to center their energy in the home. But other researchers are less convinced of this explanation. Interviews with working women, especially those from younger generations, reveal that they want to "have it all": a successful career, a happy marriage, and children (Gerson 2001; Hoffnung 2004). But when they can't have it all, they shift their preferences to what they can have. One approach, outlined earlier, is to scale back on work hours so that family goals can be satisfied (Moen 2003). A complementary tactic is to redefine career goals to harmonize with family demands. For example, an early study showed that women tend to pursue self-employment as a means to balance work and family commitments, and these types of jobs are commonly pursued after the birth of children. Such is the case for African American female entrepreneurs who establish beauty salons (Harvey 2005). But children have little effect on men's pursuit of self-employment, and men tend to use self-employment as a means to advance career objectives (Carr 1996). Interestingly, a more recent study revealed that a woman's social class may affect her reasons for self-employment. The study found that women in lower-level jobs pursue self-employment as a means of balancing work and family, but women pursuing higher-powered professional careers act like men, and when they do become self-employed, do so primarily for career reasons (Budig 2006).

In sum, socialization continues to shape men's and women's career pursuits, as well as their expectations about what they should do off the job. The tensions evident in the new economy tend to push workers to revert to the gender templates established in the old economy, with women assuming primary responsibility for the management of the domestic sphere, and men adopting roles as breadwinners. But this adoption of old gendered strategies for managing work and home occurs in a new economic and cultural context, in which most workers want and need to remain attached to the paid labor force. Liberating work in the new economy will require a serious response to the culturally based forces that push working families to adopt neo-traditional arrangements—a practice that tends to cost both men and women the opportunity to work as equals inside and outside the home (Deutsch 1999).

Interpersonal Discrimination in the Workplace

Socialization contributes to gender inequalities at work by influencing the supplies of male and female workers seeking entry into different fields. But gender also plays a role in shaping the demand for workers with specific qualities. To illustrate how this happens, imagine interviewing a series of applicants for a demanding job, a position that will require long hours of work and some travel. Your choice comes down to two candidates, Jane and John, individuals who have identical qualifications and who performed equally well in all stages of the interview process. The only difference between the candidates is that Jane is visibly pregnant, but John (obviously) is not. Which person would you hire? We posed this question to hundreds of students over the years, but only rarely has any student offered even mild support for hiring Jane. They have eloquently (and often emphatically) argued that her candidacy poses a number of concerns, including the inevitability that she will want to take time off from her job, that she will not be able to put in as long hours as John, and that she will be in no position to travel once her child is born.

All of these conclusions are based on **gender schemas,** stereotyped assumptions about women's behaviors and desires (Valian 1998). Notice how parental status becomes a master status for Jane in a way that it does not for John (who conceivably could have an infant waiting for him at home). There is a ready acceptance of the assumption that Jane would want to have reduced hours after she had her child and that the child would detract from her work. But it is also possible that she could have a partner who wants to stay home or has an excellent day care arrangement. What takes precedence in favoring

Exhibit 3.7   Instead of Hiring This Woman, Most College Students Would Hire an Equally Qualified Man. Would You?

Source: Getty Images. Reprinted with permission.

John is the emphasis on crudely constructed visions of how women and men behave in different situations, and the assumption that Jane and John would behave in a corresponding manner. This is a process termed **statistical discrimination,** wherein perceptions of group tendencies become the rationale for differential treatment of individuals. Statistical discrimination is problematic for a number of reasons, not least of which is the widespread use of **stereotypes**—often erroneous assumptions about social groups that obscure the range of behaviors and abilities present within any population.

Women are sometimes subjected to **hostile sexism,** the belief that they are inferior to men at specific tasks (Masser and Abrams 2004). A most notable recent example can be found in the comments made by former Harvard University President Lawrence Summers, who in 2005 expressed his opinion that one reason why female scientists are scarce at top research universities is because of innate differences between the sexes. When a power holder such as this accepts these beliefs, the likelihood increases that he will operate on the assumption that women are likely to fail and to view their accomplishments with skepticism (Valian 1998). For example, one study found that female scientists in the 1990s needed to be twice as productive as their "more promising" male counterparts to receive prestigious postdoctoral training opportunities (Wenneras and Wold 1997). The denial of the opportunity to succeed contributes to self-fulfilling prophecies, as workplace stratification systems lopsidedly allocate opportunities to men and then reward their successes with even more opportunities to excel.

Note that women can also be the objects of **benevolent sexism,** which operates on the assumption that they are better than men at other types of activities, such as planning social events, organizing files, or caring for children. This, in turn, prompts gatekeepers to open doors for their entry into jobs that have been traditionally defined as "women's work." The problem here is that women tend to be viewed as exceptionally qualified for work that is less rewarding than "men's work," an issue we return to shortly. Benevolent sexism can also lead employers or managers to "protect" female workers, not challenging them or involving them in difficult or dangerous tasks. The result is that women may not have as many opportunities as men to develop new skills and to demonstrate their capabilities on the job.

Overt discrimination, job discrimination on the basis of gender, was a prominent contour of the old economy. This involved the visible, conscious, and intentional decision to bar women's entry into occupations. Until the mid-20th century, women were barred from entering most professional occupational fields and it was customary for them to be paid lower wages than men, even when they performed the same work. Not until the late 1960s and early 1970s did elite colleges such as Dartmouth and Princeton admit women into many of their professional programs. Similarly, overt discrimination excluded women from well-paid jobs in the skilled trades. Entry into apprenticeship programs was difficult or impossible for women; union membership was largely restricted to men, and women who *did* succeed in entering male-dominated trades were often met with a hostile reception that made it difficult for them to "learn the ropes." Today, there are far fewer instances of overt discrimination, largely because of the **Civil Rights Act of 1964**—which prohibited employment discrimination based on sex (as well as race, color, religion, or national origin)—and the **Equal Pay Act of 1963**—which prohibited paying men and women different wages for equal work. These laws, and a series of successful lawsuits against employers, helped to change work policy and employment practices, which in turn has opened avenues for women to enter into historically male-dominated professions and industries. Overt discrimination still occurs in the new economy, but at least in comparison with the 1950s and before, the problem is less severe and women now have legal recourse.

When faced with overt discrimination, workers who know that they are being treated in an unfair manner are ostensibly in a position to level complaints or to sue their employers. Such was the case for Ramona Scott, one of 115 women who testified in a class action gender discrimination lawsuit against Wal-Mart. Scott testified that, after being passed over for promotion, her manager told her, "Men are here to make a career and women aren't . . . retail [work] is for housewives who just need to earn extra money." Stephanie Odle, an assistant manager, joined the same class action lawsuit against Wal-Mart only after she learned by chance that her male coworker earned $10,000 more a year than she did (Head 2004). Still, the likelihood that women will take advantage of their ability to sue or complain about their employer is limited because information about employee pay is usually kept closely guarded. As a result, many workers never learn the extent to which they have been treated unfairly. Even female scientists at MIT had little understanding of the extent to which they were disadvantaged in work assignments, promotions, and allocation of lab space until a group of them shared their experiences and saw patterns that otherwise were invisible (MIT 1999). Leveling complaints can also involve career costs because even when these complaints are legitimate, the victims can be perceived as troublemakers (Gallant and Cross 1993).

Women and members of minority groups can also be subject to **covert discrimination,** wherein power holders structure opportunities in a biased manner, but in ways that the employee may find difficult to detect or prove (Benokraitis 1997). A manager who is reluctant to employ or promote women, for instance, can try to increase the chances of failure by making the female employee's life exceedingly difficult. A female employee may find herself short-staffed, her hours cut, assigned unfavorable tasks, or scheduled for work at times incompatible with her needs off the job. Similarly, she may be set up for failure by being placed in situations for which she has not been properly trained or on jobs that have low odds for success (Zimmer 1986). Covert discrimination is one of the most frequently cited contributors to the existence of a **glass ceiling**—an invisible barrier that prevents female white-collar workers from rising to the highest ranks.

Discrimination is less often practiced deliberately *against* women, but rather *for* men, whom managers see as especially worthy of opportunity. Among managers, "fitting in" is of paramount importance, which influences their tendency to gravitate to, and create, gender and racially homogenous work teams (Jackall 1989; Kanter 1977). White men hold most positions of power, so they have the greatest leverage in reproducing gender and racially homogenous managerial teams (Elliott and Smith 2004). The protégé, the junior employee who looks, acts, and identifies with the (white male) boss, has a decided advantage in this culture and may be rewarded by supervisors' designing new jobs fitted to his strengths. The women excluded from old boys' clubs sometimes suspect that they have been wrongly passed over, but proving the existence of discrimination in such a culturally mediated process can be challenging (if not impossible).

An additional problem confronting women can be the existence of a **hostile work environment,** a situational context so caustic that it undermines the ability to perform jobs. Although not formally barred from working in the company of men, women can find themselves in uncomfortable situations or subjected to unwanted sexual propositions. Fully one in three women in their mid-20s has experienced unwanted touching or invasions of their personal space on the job (Uggen and Blackstone 2004). Work in such contexts can send subtle and not-so-subtle messages to women that they either do not belong or that they are primarily valued as sexual property and conquests; it can also undermine their productivity and advancement (Benokraitis 1997; Gallant and Cross 1993). Interestingly, similar reports of hostile work environments are offered by men who try to enter traditionally female-dominated professions such as clerical work or nursing (Henson and Rogers 2001). However, the uncomfortable gender environment men encounter in these situations can also produce **glass escalators** that lead them into higher-level, better-paid jobs. Male kindergarten teachers, for example, may encounter a difficult work environment (especially from hostile parents) in which questions are raised about their motives, their sexuality, and so forth. Far from harming their career prospects, this can actually encourage the transformation of these teachers into school principals, a more conventionally male role in which they acquire greater authority and higher pay (Williams 1991).

Many women are placed by their superiors in no-win situations, forced to choose whether to acquiesce to unwanted relationships, or resist and experience near-inevitable, career-damaging repercussions (MacKinnon 1998). Male-dominated workplaces tend toward masculine cultural values and behavior, and in this world, sexual jokes and horseplay can be especially problematic if there are only a few women represented. This also poses problems for sexual minorities, whose participation in sexual discussions or office romance is unwelcomed and received with hostility (Giuffre and Williams 1994; Woods and Lucas 1993).

Policing sexuality in workplaces of the new economy is problematic for a number of reasons. One concern is that many consensual romantic encounters (and marriages) are formed as the result of working together, and it is not unusual for spouses to seek work within the same organization (Astin and Milem 1997; Creamer and Associates 2001; Sweet and Moen 2004). Additionally, work cultures (accepted by both male and female employees), such as those of restaurant waitstaff, are often accepting of sexually laced coworker interplay, including pinching, joking, and even casual encounters (Williams, Giuffre, and Dellinger 1999). Other jobs, such as those of comedy writers and those in the publishing industry, require employees to develop materials that some may find objectionable, and this in turn may influence the gender composition of fields (Exhibit 3.8). Attempting to constrain discussions or eliminate potentially offensive materials could ultimately undermine the prospect of actually being able to do this work (Dellinger and Williams 2002). In the service economy, selling sexual titillation (such as at a Hooters restaurant) is part and parcel of some jobs.

How do most workers respond when they feel they are being discriminated against or find themselves in hostile work environments? Among those in low-level jobs, such as employees in the fast-food industry or

Exhibit 3.8   *Daily Show* Writers and Producers Receive Their Emmys in 2005: Why Are There So Few Women?

Source: © Mario Anzuoni/Reuters/Corbis. Reprinted with permission.

retail sector, the most common response is to tolerate the situation, and when that fails, to try to find work elsewhere (Tucker 1993). But for women in higher-level positions, harassment or inequitable treatment on the job can confront them with a classic "catch-22." Even in situations where the worker can document unfair treatment, by relying on channels (e.g., complaining to her boss' supervisor), she reveals a failure to "fit in" or "get along" (Gallant and Cross 1993). Ultimately the boundaries of acceptable conduct in the new economy will be played out in the courts, as well as in the ongoing negotiation of informally established workplace cultures.

Structural Dimensions of Gender Discrimination

Thus far, we have considered the problem of gender discrimination largely by focusing on how individuals respond to one another on an interpersonal level. If gender inequality were simply a matter of interpersonal discrimination, solutions would involve changing people through initiatives such as training seminars or leveling fines against those who act irresponsibly. However, even if these initiatives were successful, deeper structural forces that reinforce gender inequities in the new economy would remain. In this section, we focus on two institutionalized practices that limit women's abilities to compete in the modern workplace—the ways in which women's work and men's work are valued, and the ways in which job designs conflict with gender scripts.

The Devaluation of Women's Work

Consider the differences between two low-status occupations, one of which is primarily occupied by men, the other by women. The job descriptions in Exhibit 3.9 are quoted from the *Occupational Outlook Handbook* (Bureau of Labor Statistics 2006b), a handy source of information about the nature of different jobs and the economic returns workers receive for their efforts.

Exhibit 3.9   Two Different Job Descriptions: Why Does One Job Pay Less Than the Other?

Refuse and Recyclable Material Collectors

Gather refuse and recyclables from homes and businesses into their truck for transport to a dump, landfill, or recycling center. They lift and empty garbage cans or recycling bins by hand or operate a hydraulic lift truck that picks up and empties dumpsters. They work along scheduled routes.
Median pay in 2004, $12.38/hour.

Child Care Workers

Nurture and care for children who have not yet entered formal schooling and also work with older children in before- and after-school situations. These workers play an important role in a child's development by caring for the child when parents are at work or away for other reasons. In addition to attending to children's basic needs, child care workers organize activities that stimulate children's physical, emotional, intellectual, and social growth. They help children explore individual interests, develop talents and independence, build self-esteem, and learn how to get along with others.
Median pay in 2004, $8.06/hour.

Source: Bureau of Labor Statistics (2006b)

Why does a garbage collector earn $4.32 per hour more than a day care worker? One could suggest a variety of reasons, but none of them eliminates the fact that day care work is more challenging and requires greater skill. Disparities such as these have led feminist scholars to consider the **comparable worth** of different types of jobs. These scholars argue that wage disparities are not simply the result of labor supply and demand but rather, are intrinsically tied to the valuation of men's and women's work. Of course, one should use caution in generalizing from a selective comparison of two specific occupations. To address this concern, Paula England and her colleagues developed a variety of statistical approaches to document how comparable worth affects women's earnings, considering not so much who is working but rather, the jobs that are associated with men's and women's roles in society. England's statistical models show that if a man with the same level of education moved from an occupation that was entirely composed of men to an occupation that was entirely composed of women, his earnings would fall by more than $2 per hour. Jobs that require care work (which is culturally associated with women's responsibilities) pay 5% to 10% less than do jobs that do not, even if other relevant factors are considered (e.g., jobholders' education, years on the job, supervisory responsibilities). It is hard not to conclude that "women's work" is systematically undervalued in the paid economy (Aman and England 1997; England, Budig, and Folbre 2002; Karlin, England, and Richardson 2002).

There are often marked disparities in the comparable worth of *tasks* within jobs as well. Most occupations require workers to perform a wide range of activities, all of which are central to organizational success. New employees need to be trained, notes need to be taken, interpersonal conflicts need to be resolved, and a variety of other personnel and technological concerns need to be addressed on a day-to-day basis. Men and women who are employed in the same type of job may be directed to perform different aspects of that work. For example, female professors tend to be assigned a disproportionate share of student advisees, which in turn frees their male colleagues to spend more time on more highly valued research activities. Although colleges and universities give lip service to the importance of advising students (and cannot succeed without this work), the time spent on this work can hurt, rather than enhance, an employee's performance review. In the end, the tasks that men tend to be assigned carry greater weight in decisions about promotions and pay increases (Fletcher 2001).

How Job Designs Discriminate

Nearly all police departments require patrol officer candidates to demonstrate physical aptitudes for the job and rely on tests that commonly include running and assessments of strength. On the surface, tests of agility and strength appear to be gender-neutral because the same standards are applied to all candidates. But when one looks at the effects, and how fitness exams tend systematically to screen out female applicants, structural biases in their construction and application are revealed. At one police department in Pennsylvania, for example, only 1 in 10 (12%) female applicants successfully passed the running test, compared with more than half (60%) of the men (Brooks 2001). On measures of strength, women also fare poorly. The average woman can only perform sit-ups and push-ups at about 75% the capacity of men, and the average grip strength for women is only 57% of that of men (Shephard and Bonneau 2002). These differences are real, but one must ask how often a police officer will actually need to perform push-ups or run long distances in the line of duty. One also wonders how much grip strength it actually takes to fire a pistol, and why tests of strength, such as these, are on fitness exams. Nevertheless, these are the standards commonly used to screen applicants.

Part of a patrol officer's job requires physical work, and applicants should not be exempt from those responsibilities. However, it should be the case that the tests used to assess applicants for a job are *relevant* to the job. Most day-to-day police work involves mundane rides in patrol cars, writing up reports, processing people, and handing out tickets. When force is needed, it is commonly performed as a team effort rather than being performed by lone officers. Additionally, it is interesting to note the physical tests are used to determine who can *enter* into police jobs, but not who can *keep* them.

Selective testing techniques also tend to ignore the skills that women are especially likely to possess. For example, only about half of all police departments include situational tests, wherein a simulated real-life encounter is constructed and recruits are judged on their ability to handle routine traffic stops or more dangerous encounters (Shephard and Bonneau 2002). In these simulation exams, the ability to read emotional states and intentions is the needed skill, a strength more commonly associated with feminine qualities than masculine attributes (Martin 1982). Arguably, interpersonal tests are more important than physical tests because the ability to read a person's motivations and emotional state, and respond accordingly, could help officers avoid having to use physical force in the first place. Because these tests are not as widely used, the likelihood that candidates with poor interpersonal skills will make it through the screening process is increased, which favors the candidacy of underqualified male applicants. Also, note that the work of police supervisors or detectives almost never requires the use of force, but the career path to those positions is through patrol jobs, which in turn requires demonstration of physical prowess.

Structural forms of discrimination are built into the very design of work, and therein lies the problem, because most jobs and professional expectations have been designed with men in mind. For example, to become a pilot in the U.S. Air Force, one needs to be 64 to 77 inches tall, a standard that is squarely in the midrange for American men, but that screens half of all women from eligibility.[1] Here we observe something different than the police exams, in that height criteria are not arbitrary or irrationally selected. Pilots need to be able to reach the levers and buttons that surround them on all sides of their cockpit seats. In this case, the structural discrimination occurred in the design of the plane itself, which was tailored to fit the typical man, not a typical woman.

The legal scholar Joan Williams (2000) identified a number of interesting cases such as these in her book *Unbending Gender,* which examined the templates used to construct job expectations. She concluded that the current structure of work itself can be discriminatory because it uses the image of the male worker as the metric for determining the characteristics of the ideal worker. Ideal workers, in today's economy, are people who can put in long hours, uninterrupted, throughout their careers. They are available when the company needs them, can work late when called upon, can spend weekends in the office, and will not have any prolonged absences from the job. Their performance will not be hindered by sickness, aging parents, children, or pregnancy. These expectations are largely insensitive to the *normal* life courses of women and their roles off the job. In a word, ideal workers are men (see also Moen and Roehling 2005).

As another example, consider how inhospitable most workplaces are to breastfeeding—a normal part of women's lives, but not men's. Exhibit 3.10

Exhibit 3.10   Health Benefits to Breastfeeding

Benefits to Children	Benefits to Mothers
Less urinary tract infection	Less breast cancer
Less respiratory infection	Less ovarian cancer
Less diarrhea	Less osteoporosis
Less allergic disease	Earlier weight loss
Less middle ear infection	Enhanced self-esteem
Less bacterial meningitis	Enhanced infant bonding
Less botulism	Enhanced feelings of success
Less bacteremia	
Less gastrointestinal infection	
Less atopic eczema	

Source: American Academy of Pediatrics (2005)

reveals the numerous health benefits of breastfeeding, for both children and nursing mothers. The American Academy of Pediatrics (2005) recommends that newborn children be exclusively fed breast milk for the first 6 months of life, and then be provided with breast milk along with other food until at least 12 months of age. But fewer than one in four American women who work full time nurse their children to the 6-month marker, a much smaller percentage compared with new mothers who are out of the labor force (Galtry 2000, 2002).

The breastfeeding problem can be attributed to the fact that most working women cannot afford temporary leaves from their jobs and that newborn children are not welcomed in most workplaces. But there also exists an underutilized technological solution that can help women stay strongly attached to jobs. Breast pumps (which used to be bulky and somewhat noisy appliances) are now quiet and small enough to be carried like briefcases. These devices enable nursing mothers to express milk while away from their children and then store it for later use. Using these machines only requires privacy, access to a refrigerator, and the provision of brief breaks (20–30 minutes a few times a day). These are modest resources to provide, but they remain unavailable to most workers.

In sum, a focus on job designs and the standards used to evaluate employees reveals structural considerations that disadvantage women in the paid labor force. Changing this requires rethinking taken-for-granted assumptions about what a diverse workforce can reasonably contribute to their jobs. It requires new human resource policies and practices to generate interest in jobs, screen applicants, design work spaces, construct work schedules, and evaluate employee performance. The first step in this process is to rethink who the "ideal worker" is, thereby bringing workplace policy and practice into line with what is reasonable to expect from today's diverse workforce.

Strategies to Bridge the Care Gaps: International Comparisons

The increased participation of women in the paid labor force has contributed to resource gaps in the ability to care for children and aging parents. This problem is going to escalate in future decades as the aging baby boom generation transitions into retirement and then into old age. As this happens, ever-larger proportions of the workforce will find themselves sandwiched between the need to work, care for their parents, and care for their children (Neal and Hammer 2006; Sarkisian and Gerstel 2004). Here we consider two core policy considerations—family care and family leave—and how those policies in the United States compare with those in other advanced societies.

America's approach to handling day care is a hodgepodge of stopgap approaches that create divergent experiences among different classes in the workforce. For workers at the top of the class structure, such as Erika, the stay-at-home mother discussed at the beginning of this chapter, high incomes afford not only the luxury to stay home, but also the resources to employ others to cook, clean, and at times watch children. Professional women with demanding jobs can purchase care privately, but often at considerable expense, and rely on a variety of prepackaged care items, such as prepared foods. These strategies work, but eat into couples' incomes, stall careers, and create time binds that strain marriages. For workers in jobs that pay low wages, the strains are far greater and solutions are even less satisfactory.[2] Many parents in working-class families deliberately work alternating shifts, so that someone is home at all times. This can work, but it has obvious negative effects on marriages and can create high levels of stress in people's lives (Presser 2003). It is not unusual for children in low-income families to be left to care for themselves (Heymann 2000).

A full understanding of the problem with the American system of day care requires considering the ways in which care needs are connected with service provision. Today, America's day care system operates on the backs of poorly paid domestic and child care workers, who are almost exclusively women and disproportionately members of minority groups or recent immigrants. The affordable labor they provide makes it possible for professional women to devote time to their jobs, revealing that the ability of some women to integrate themselves into the new economy depends on the exploitation of other women (Mohanty 2003).

The quality of and access to day care in American society pales in comparison with many countries in Western Europe, which have implemented national child care programs that provide high subsidies to fund publicly financed child care centers, which in turn promotes high use (Gornick and Meyers 2003; Pettit and Hook 2005). Because most child care workers in Sweden are well paid and have university degrees, virtually all working parents can go to their jobs assured that their children are receiving high-quality care (Morgan 2005). American culture embraces the notion that children are best raised in the warm arms of their mothers and considers day care providers to be comparatively "cold." This perception is not shared in Denmark, a country with universal child care supports. Danes believe that children served in quality state-sponsored centers receive "warm" care that helps them grow and mature, and it is not considered inferior to the care received in the home (Kremer 2006b).

Subsidizing care centers is one means of promoting family-responsive policies. But this does not address the fundamental problem that most care work is uncompensated and performed by family members (Crittenden

2001). Here another social experiment is enlightening—the provision of direct payments to individuals to purchase care assistance. In the United States, professional home health care assistance, such as in-home nursing, is available only to those who have financial resources. Those lacking resources commonly go without care or rely on family members to provide this care (which in turn can interfere with their work). In Denmark, those needing home assistance are given a "Personal Budget" that they can use to purchase the assistance of their own choosing. Interestingly, 64% of the people employed by those controlling their own Personal Budgets are the spouses, parents, or adult children of the person purchasing care. In essence, this system enables family members financially to compensate each other for providing care work, and legitimates their taking time out from the paid economy. However, this system is not without its own problems because role conflicts can result when family members employ one another. It also has undermined the power of professional groups to define reasonable care standards (Kremer 2006a).

There have been some advances in the U.S. approach to work–family policy, although they are modest by international comparison. One of these occurred in 1993, with the passage of the **Family and Medical Leave Act (FMLA)**.[3] This legislation provides workers with the right to take 12 weeks of leave from their jobs to care for newborn and newly adopted children or to assist other family members in need. By recognizing workers' need to take "time-outs" to tend to family matters, the FMLA was a tremendously important step in helping address existing care deficits. But it has not helped all workers. Because the leave allowed is relatively short (12 weeks), the amount of care work that can be performed is limited. Employees who have worked for their employers for less than 1 year, those who work part time (less than 25 hours/week), and those who work for smaller employers (with fewer than 50 workers) are not covered by the FMLA. This works to the disadvantage of low-income workers, who are much more likely to be employed in these smaller enterprises, to have less stable job histories, and to work part time. Most importantly, because the leave is unpaid, it is useful only to those workers who can afford to take time away from their jobs. Ironically, the result is that those who most need to take time away from their jobs—those who cannot afford to pay for care services—are excluded from reaping the benefits of the FMLA (Gerstel and McGonanagle 1999).

The FMLA is family leave American-style; it assumes that most families have a breadwinner who can remain in the labor force to provide for family economic needs. For those at the lower end of the economic spectrum, family leave can be considered a right, but not an option. Exhibit 3.11

Exhibit 3.11   Family Leave Entitlements in Developed Countries

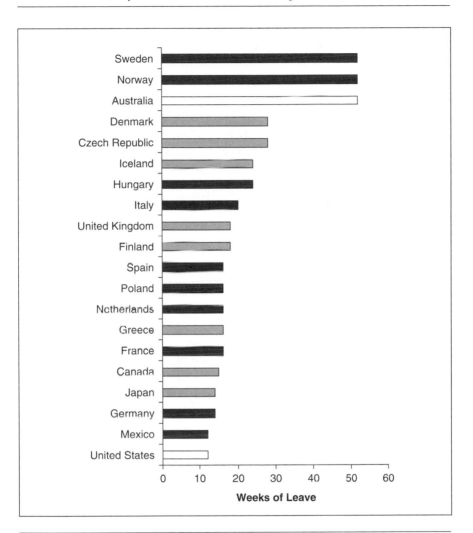

Source: Kamerman (2000)

▬ Black bars represent paid leave (80% or higher)

▨ Grey bars represent partially paid leave (most typically 50%)

▢ White bars represent unpaid leave

shows that the American approach is vastly inferior to the leave policies adopted in nearly all other developed nations. For example, Sweden and Norway offer new parents an entire year off after the birth of a child, and these workers receive 80% of their base pay during that time. Like most of the countries that have paid parental leave, it is not the employer who pays (as Americans commonly assume would be the case). Rather, the government provides stipends (supported through higher taxes) that replace the wages new parents would have earned from their jobs. Although most Western European countries have developed more family-responsive policies, important cultural variations have also influenced the forms these policies take. Some societies seek to emancipate women from the home and dismantle traditional gender roles (e.g., Finland), whereas others seek to facilitate the fulfillment of traditional gender roles with some integration of women into the paid labor force (e.g., the Netherlands and Germany) (Pfau-Effinger 2004).

Sweden offers an interesting case study of the possibilities for policy change, as well as cautionary findings on the impact family-responsive legislation can have on men's and women's behavior inside and outside the workplace. In the early 1970s, the Swedish government formed a commission to study family needs. Its findings pushed the nation to enact and actively promote family leave with legislation that was initially designed to encourage both mothers and fathers to be involved in the nurturing of children. It soon became clear that men, on the whole, were unwilling to make use of parental leave benefits. In response, the government initiated legislation and public relations campaigns to promote a "daddy month"—an incentive specifically directed to encourage men to take paid time out from their jobs to tend to their children. Interestingly, the impetus for this initiative came first from women's groups, who saw rebalancing family divisions of labor as a critical condition for increasing gender equality; later men became highly involved and helped win passage of the legislation (Bergman and Hobson 2002).

Sweden's experiments in family-responsive programs revealed an irony. Although paid family leave facilitated women's participation in the labor force and helped them care for young children, it also *exacerbated* the level of gender occupational segregation and the male–female earnings gap. The reason for this dynamic is that paid family leave is almost exclusively used by women (Morgan and Zippel 2003). When employers know that men are unlikely to request time away from the job, but that women will be likely to make that request, clear incentives are created to discriminate against female workers. As a result, countries with the most family-friendly legislation (such as Sweden) have even higher levels of sex segregation than the United States does, and women are less likely to advance to the highest levels in the paid labor force (Mandel and Semyonov 2004, 2006). This does not mean that one should do away with family-responsive policies. On the contrary, these need to be expanded—especially in the United States—along with other initiatives to redefine "women's work." But cultural initiatives also need to direct men to be more deeply engaged in work in and around the home. If men demanded and used family leave as frequently as women do (as is usually their right), the disincentives for hiring women would diminish.

At a time when many European countries have been expanding their supports for working families, the United States enacted regressive policies directed at those at the bottom of the class structure. Ironically, as the need to care for disadvantaged children increased, welfare reform has undermined the prospects for poor single mothers to stay home with their children. The Personal Responsibility and Work Opportunity Reconciliation Act of 1996 reworked the previous major welfare program, called **Aid to Families With Dependent Children (AFDC)** into the current program called **Temporary Assistance for Needy Families (TANF)**. The AFDC system provided monthly welfare checks to poor families as a substitute for attachment to the labor force. Political leaders came to believe that this approach was promoting a culture of dependency, one in which poor people accepted "welfare as a way of life" (Murray 1995). In its stead, TANF was designed to limit support to a lifetime maximum of 5 years (states are free to set the limit lower, and some have done so), as well as to create an expectation that recipients work while receiving assistance (and continue to work after the assistance ends). In this way, TANF was designed to get the poor into the labor force. But it did so with inadequate provisions to enable poor families (disproportionately single mothers) to manage the care of their children (Hays 2003). Studies of these families on the fault line indicate that they are typically placed in jobs that pay at or near the minimum wage, that these jobs seldom have any prospects for upward mobility, that care responsibilities interfere with women's abilities to keep these jobs, and that the lack of child care resources forces them to place children in situations untenable by middle-class standards (Crouter and Booth 2004). So, whereas many societies in Europe have enabled women to take sabbaticals from their jobs, America introduced initiatives to require the participation of the poor in jobs that pay little and offer few opportunities for advancement. Time will tell what impact this has on the next generation, although it is reasonable to predict that forcing poor women to

take low-wage jobs will do little to alter existing gendered wage disparities and will maintain or even exacerbate racial disparities as well.

Conclusion

In this chapter, we surveyed the contours of gender inequality in the workplace, mapped the shape of these inequities, and considered the reasons for
their occurrence. We showed that even though nearly half of the American
workforce is female, only a small fraction of these women rise to the top
and become corporate or government leaders. Instead, they are disproportionately funneled into lower-paying jobs and face greater prospects than
men of having their careers dislodged midstream. We also showed that
women perform substantial amounts of work that is economically unrecog-
nized or undervalued. Although interpersonal discrimination remains an
important consideration, America needs to remedy deeper structural problems that differentially affect women's and men's careers. To be successful
in today's workforce requires molding oneself into jobs designed for men
who had stay-at-home wives.

The chasms that separate men's and women's careers are created by a
variety of forces that have the consequence of creating disadvantages for
women that accumulate over their life courses. Women are coached early
on to have different aspirations than men, and men and women are encouraged to accept—as a given—that "women's work" is of lesser economic
value. If this work is undervalued, men will not seek it, with the result that
large numbers of women will remain trapped in low-paying occupational
ghettos. There have been advances in encouraging girls to aspire to enter
"men's" careers. But there is a crying need to teach boys (and men) the
rewards that can be attained in the performance of "women's work," in the
home and in the paid economy, and to alter existing practices that systematically undervalue care work.

Although not a panacea or without problems and concerns, America's
adjustment to the new economy can benefit from considering the far bolder
European social experiments that advance opportunities to care for children
and other needy relatives. Ultimately the future structure of the new economy
will reflect how society responds to the question of gender inequality and its
level of commitment to dismantling the gendered divisions of labor both
inside and outside the home.

Notes

1. The average height of American men is 69.2 inches, for women, 63.8 inches.

2. Parents in the United States can take a pretax deduction for child care expenses, but this policy does little to help low-income families who commonly pay little in taxes to begin with.

3. In California, legislation was passed in 2004 mandating that employees be eligible to receive as much as half pay for a period of 6 weeks to care for a sick or injured family member, or following birth, adoption, or foster care placement. Unlike the FMLA, there is no restriction on organizational size or tenure with the employer. Time will tell if the country follows California's lead, one that was met by considerable opposition from business groups (Nowicki 2003).

4

Race, Ethnicity, and Work

Legacies of the Past, Problems in the Present

J amal, the young African American fast-food employee introduced early in this book, has the odds stacked against him. He grew up in a family that lacked resources, and in a community that had few opportunities. But one wonders what factors caused his current situation, his tenuous hold on work in the new economy. Are his problems simply the result of his (and his parents') choices? Or are they the result of his lower-class social origins? No doubt both played a role. However, in this chapter, we will argue that race and ethnicity continue to matter. Race, ethnicity, and the new economy are intertwined, leaving a generation of young, lower-class minority workers at the margins. And even among the successful, race continues to influence how high one can rise.

In this chapter, we offer an overview of race, ethnicity, and work, looking both to the past for the origins of contemporary problems, and to the future challenge of creating a fully integrated workforce. We first consider race and ethnicity from a historical perspective to provide a context for considering the ways in which racial/ethnic exclusion manifested themselves in the old economy. Although patterns of exclusion have changed in some ways, they continue to influence the allocation of the resources needed to prepare for work, as well as access to jobs themselves. As a result, members of some racial and ethnic minorities are at a distinct disadvantage in today's workplace and labor market. These observations lead us to several critical policy issues concerning work, race, and ethnicity in the 21st century, including immigration control and affirmative action programs.

Histories of Race, Ethnicity, and Work

"Race" and "ethnicity" have been vitally important issues in the history of the United States. What is meant by these two terms, however, is not always entirely clear. Complicating the issue further, racial categories have changed dramatically over time. For example, Americans now refer to all persons of European descent as being "white," whereas early 20th century Americans would have categorized them into many different racial groups. Sociologists and others have increasingly come to define race as a social construct, in which physical differences between groups (such as skin color) are assumed by members of society to mean that those groups inherently differ in many other ways (ability, intelligence, etc.). The term "ethnicity" (or "ethnic groups") is generally used to describe the cultural differences (language, diet, costume, etc.) between groups (for example, between Americans of Polish and Irish descent) (Omi and Winant 1986).[1]

African American Exceptionality

One question commonly posed is why African Americans, who have been in America nearly as long as white Americans (and much longer than many other ethnic groups) are economically so far behind. One explanation is that, unlike virtually all other immigrant groups, most people of African descent were brought forcibly to America, severed from family ties, and sold into bondage. In early America, African Americans were, in most instances, forbidden to hold property or to advance their positions. The dominant ethnic group, White Anglo Saxon Protestants (WASPs), accrued wealth and later came to control the economic and political fortunes of the 19th century, but African Americans were left behind. Although American history books mark the resolution of the Civil War as the formal end to slavery, it did remarkably little to change the day-to-day work experiences of African Americans in the South, where most newly freed African Americans remained propertyless, sharecropped the fields for their old masters, and accrued debt rather than wealth.

Although a significant number of African Americans worked outside of agriculture even under slavery, it was difficult for them to find work in industry in the post–Civil War South. Some did find employment, but even those with significant skills were typically relegated to work as unskilled laborers. Whites and most trade unions (especially unions of skilled workers) devoted considerable energy to keeping African American workers out. And the racial hostilities that permeated American culture were strategically exploited by employers in some parts of the country, using African Americans as strikebreakers or threatening to replace low-wage white workers with African Americans unless workers acquiesced to management's terms. African American women were much more likely than were their white counterparts to work in the post–Civil War period, but they were largely kept out of industrial employment. Most found work in various menial service occupations, such as clothes washing (Gerstel and Sarkisian 2006; Harris 1982).

Around the time of World War I, large numbers of African Americans (along with rural whites from Appalachia) traveled from the South to the North to find work in the rapidly industrializing cities. Even though male African American migrants to the North did succeed in finding work in industry, most remained confined to unskilled jobs at the bottom of the employment ladder and were generally less likely to be employed than were new immigrants coming from Eastern and Western Europe. African American women had far less success entering industry than did African American men, so they continued to work in low-level service occupations. In the South, as well as in the North, a stable African American middle class emerged, but the black–white division was so strong that both groups formed separate castes within a supposedly free society, separated into distinct communities, and shared few social ties or common experiences (Myrdal 1995 [1954]).

In general, throughout the period of rapid industrialization, most African American workers, whether they were in the South or the North, found themselves excluded from all but the least desirable forms of employment, far more exposed to unemployment in bad economic times, and prevented from acquiring the education or training that might gain them access to better-paying jobs. But there was some progress. Some unions, for example, particularly those that organized all categories of workers, not just skilled workers, organized across racial lines. And, during World War II, because of labor shortages, some African American workers moved up, temporarily, into better jobs. However, once the war was over, many of these gains evaporated, leaving African Americans in much the same condition they had been in before (Harris 1982; Honey 1999).

In the latter half of the 20th century, many aspects of race in American society were changed. Key pieces of civil rights legislation, such as the **Equal**

Pay Act of 1963 and the **Civil Rights Act of 1964,** established the legal standard that employers should not exclude anyone from employment on the basis of race or ethnicity. Civil rights legislation also put an end to legalized segregation in the South. Evidence also shows that overt hostility to African Americans has been on the wane and that the United States has been becoming a more racially tolerant society (Schuman, Steeh, Bobo, and Krysan 1998).

Does this mean that race no longer matters? Some, like the highly influential sociologist William Julius Wilson (1978) argued that race has declined in its significance. He observed that African Americans who had moved out of the central cities, who had entered college, and who had developed marketable skills stood reasonable chances of making it in the rapidly changing economy. If African Americans continued to be economically disadvantaged, he argued, it was not the result of race but of *class,* that is, of the declining opportunities for well-paid employment for those with limited education and skills.

Others disagreed, finding evidence that race still mattered. Even though legalized discrimination had been abolished, and overt forms of prejudice weakened, the United States is still not a color-blind society, and various forms of "subtle discrimination" continue to affect all aspects of social life, including workplaces (Feagin 1991). As will be made clear in this chapter, good evidence supports *both* of these assessments.

The Immigrant Experience

African American workers faced particularly well-organized and harsh systems of oppression, but they were not alone in encountering obstacles to employment in the industrial United States. Americans frequently succumbed to "nativist" sentiments, and many immigrant groups found themselves excluded from certain kinds of employment ("Irish need not apply"), trapped in poorly paid ethnic economies (e.g., the Jews in the garment trades), exploited by unscrupulous employers, and suspected of radical or anti-American sentiments. In the factories of the late 19th century, jobs were distributed to ethnically homogenous work teams, and favoritism to one's own ethnic group was considered a normal way of distributing work (Jacoby 1991). Various stakeholders of the time produced a variety of documents demonstrating the dangers these new immigrants posed to American society (Gusfield 1963; Thomas and Znaniecki 1958). As Exhibit 4.1 shows, these hostilities were predated by concerns raised in the mid-19th century about Asian immigrants. This cartoon not so subtly suggests that immigrants, and their foreign ways, had the potential to swallow up American values and liberties.

Exhibit 4.1   An Editorial Showing Fear of Immigrants (Circa 1860)

Source: © Corbis. Reprinted with permission.

During the 20th century, some hostilities faded as immigrants gained citizenship and assimilated. For the most part, immigrant families who had seemed foreign and threatening were amalgamated into the new economy, improving their fortunes with the expanding U.S. economy. But suspicion of immigrants remained strong, as evidenced by the passage of highly restrictive legislation that established strict quotas on immigration from virtually all parts of the world (Alba and Nee 2003). Here we offer but a few examples:

- Male Chinese workers were imported to the United States, largely to help build the transcontinental railroad, but they were not allowed to bring along their families or become citizens. Once the railroad was completed, Americans reacted to the arrival of immigrants from China and Japan by pressing for legislation to keep them out. These efforts were successful: Chinese immigration was stopped with the 1882 Chinese Exclusion Act and Japanese immigration with the Gentleman's Agreement of 1907 and the Immigration Act of 1924.

- Japanese immigrants to the United States (Issei) succeeded as farmers in Western states and hoped that their children (Nisei) would gain acceptance into the American mainstream. However, despite doing well in school, few managed to find jobs outside the ethnic labor market before World War II. Isolated and disliked, Japanese Americans were easy targets for the internment policy of World War II (Takaki 1993).

- The Bracero program, established to respond to labor shortages during World War II, established work opportunities for immigrants from Mexico. As "guest workers," Mexicans were allowed to enter the United States temporarily to work and then return to Mexico, but the program did not permit long-term residency and restricted their employment to low-level jobs American employers found difficult to fill (Alba and Nee 2003).

Changes in immigration law in the post–World War II period put an end to the exclusionary laws affecting Chinese and Japanese immigration and eliminated the immigration quotas, with their "racial" overtones that had shaped immigration policy since the 1920s. The Bracero program, too, was ended in the mid-1960s, eliminating the special category of "guest worker," to which many Mexican migrants had been confined. Recent immigration law reforms opened the door for increased levels of immigration. Although the percentage of Americans who were foreign-born was much higher in the early 20th century, levels of immigration to the United States in the late 20th and early 21st century rivaled those of earlier periods, with more than a million new *legal* permanent residents being admitted in most years (Department of Homeland Security 2005).

As levels of immigration rose, a sense developed that the most recent immigration waves are different. New immigration rules gave priority to those with educational credentials, which influenced the types of immigrants entering the country and their ability to enter the good jobs being cultivated in the new economy. As we discuss later, this dynamic can help explain why some recent immigrant groups have succeeded in ways that resident African American groups have not. However, because priority was also given to immigrants with close family in the United States, many legal immigrants continue to arrive with limited education and skills (Alba and Nee 2003). Many are concerned that these immigrants will become a burden on society or that they will bid down wages and displace more expensive American-born workers, an issue we will discuss at greater length later in this chapter. Particular attention was paid to cultural issues, as Latino migrants were (falsely, as it turns out) believed to be resistant to learning English (Portes and Rumbaut 2001). Some also felt that, in an age of easy electronic communication and cheap travel, immigrants, particularly from nearby Mexico, would remain tied to their cultures of origin and would resist "Americanization" (Alba and Nee 2003)

In this context, hostility to immigrants has reemerged, as evidenced by attempts to build a fence between the U.S.–Mexican border and the emergence of vigilante groups such as the Minute Men, who patrol for illegal immigrants. The concerns expressed in Exhibit 4.1 are echoed by the thesis of politician/commentator Patrick Buchanan's (2006) book *State of Emergency*. The only difference is that, instead of focusing concern on the Chinese, this book directs hostility toward Mexican immigrants.

The Magnitude of Racial Inequality in the New Economy

It is hard to argue that race and ethnicity are no longer major factors in determining one's fate in the workplace after considering the extent of the differences between Latino, African American, and white American incomes. One way of measuring these disparities is to examine per capita earnings, that is, the total of all earnings within a racial group divided by total group membership. Exhibit 4.2 reveals that average incomes for men with incomes in the three major racial groups have risen only slightly during the past two decades and that the gap between African American and white men's incomes has narrowed slightly.[2] Exhibit 4.3 shows that averages for women with incomes have risen more rapidly (although women continue to trail men by significant margins); the racial divides that separate white women from Latina and African American women are far smaller than those that separate men. However, these exhibits also show the continued existence of income gaps between whites, African Americans, and Latinos that remain almost as wide in 2003 as they were in 1980.[3]

The large gaps in men's incomes are only part of the story. Another important question is whether the jobs obtained by African Americans and Latinos are sufficient to bring families to a reasonable standard of living. Answering this requires shifting analysis from individuals to families, and considering all the sources of income a family possesses. Family structures vary across social groups, with significant consequences for their earnings. For instance, African American families are far more likely than are white families to be single-parent families relying on a single income. Latino and

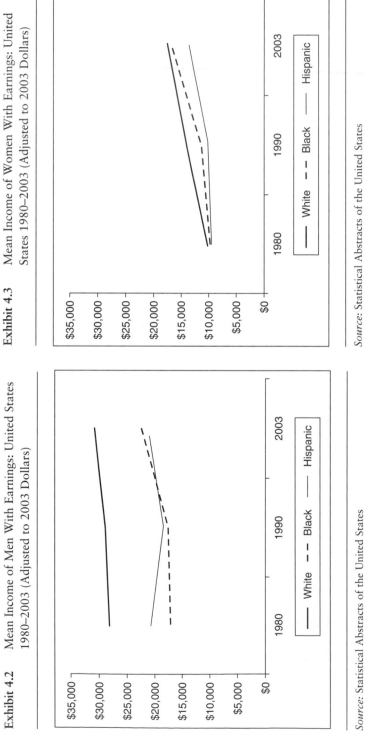

Exhibit 4.2 Mean Income of Men With Earnings: United States 1980–2003 (Adjusted to 2003 Dollars)

Source: Statistical Abstracts of the United States

Exhibit 4.3 Mean Income of Women With Earnings: United States 1980–2003 (Adjusted to 2003 Dollars)

Source: Statistical Abstracts of the United States

Asian American families, on the other hand, are much more similar in structure to white families than they are to African American families; having
two adult wage earners in the household may help compensate for individuals' being trapped in low-wage jobs.

To deal with these complexities, we compare the proportions of *families*
living below the poverty level in the four major ethnic groups. As Exhibit 4.4
shows, African American and Latino families are three times more likely
than are white families to be living below the poverty line. This reveals that
race and ethnicity are closely tied to the ability to find the types of jobs that
enable families to afford reasonable standards of living. For one in five African
American or Latino families, these opportunities are nonexistent.

Exhibit 4.4   Percentages of Families Living Below the Poverty Line:
United States 2003

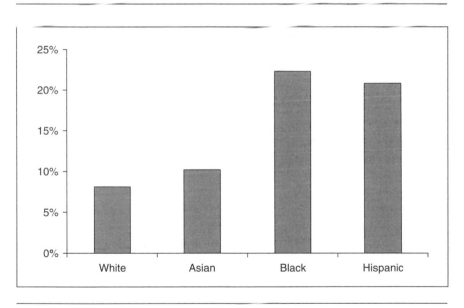

Source: Statistical Abstracts of the United States

The existence of a significant income gap between whites and African
Americans or Latinos has fueled speculation that a racially and ethnically dis-
tinctive "underclass" is emerging in the United States. This argument was
first made about African Americans, whose higher rates of poverty and bad
employment prospects are an old and (as the earlier data indicate) continuing
story (Jencks and Peterson 1991). More recently, scholars have speculated that
a similar phenomenon is developing among Latinos. These scholars argue that,
unlike earlier European immigrants, Latinos are not rapidly assimilating into

the American economy and moving their way up the economic ladder. Instead, a process of "segmented assimilation" is occurring, in which Latinos become part of the American economy, but in roles much like lower-class African Americans, that is, in poorly paid, unstable, secondary labor market jobs (Portes and Zhou 1993). Not everyone agrees with this view; some contend that Latinos *are* gradually making their way up the ladder although their progress is slowed by the shrinkage of factory employment (the traditional route for upward mobility among immigrants) (Alba and Nee 2003). But others argue that Latinos are experiencing what they call "working-class assimilation," that is, they are moving up the ladder, but not as fast as they would if they were not members of an identifiable minority group (Waldinger, Lim, and Cort 2007). Though not fully agreeing on the pace of assimilation, all these studies conclude that being Latino has a significant negative effect on the ability to find good, secure, well-paying jobs.

Thus, we are left with a puzzle. How is it possible that, in a society where the legal barriers to the employment of racial and ethnic minorities have been removed, where there is evidence that some cultural attitudes have changed, where the belief is that equal opportunity exists, and where examples of successful minorities abound—how is it possible that African Americans and Latinos lag so far behind whites and people of Asian descent?

In the sections that follow, we argue that the reasons for racial and ethnic inequalities in the workplace are tied to three overarching sources of inequality and that each of these sources of inequality involves persistent problems from the old economy and emerging problems linked to the new one. One source is the ability of parents to prepare their children for the world of work, a process that requires intergenerational transmission of resources. Another source involves aspects of social structure, such as the ways work opportunities, race, and geography intertwine. Finally, interpersonal dynamics play a role as well, as members of different ethnic groups face different prospects of being perceived and treated equitably in the workplace. As we discuss later, these separate explanations—resources, opportunity, and discrimination—work together in a cumulative manner to fashion racial/ethnic disadvantages in the new economy. We also show that ethnicity works to facilitate the formation of social connections that help members of some ethnic groups adapt to the new opportunity structures.

Intergenerational Transmission of Resources

For most workers in today's economy, a successful career requires strategic investments of time, energy, and financial resources. In this section, we consider four different types of resources that are passed through intergenerational relations. These resources constitute **capital**—assets that can be used to enhance future career prospects (Bourdieu 1986). One type of capital is economic—income and property that can be devoted to education or entrepreneurial endeavors. But prospects for success can also be elevated through the development of marketable skills (human capital), network ties (social capital), and the ability to interact comfortably with others who have resources (cultural capital).

Race, Ethnicity, and Economic Capital

In comparison with whites, African Americans and Latinos have less **economic capital**—financial resources essential to pursuing a college education or the start-up funds needed for entrepreneurial endeavors. We have just seen that income is unequally distributed; so is wealth, the sum total of all possessions. Wealth is difficult to measure because it is not formally tracked in the way that income is assessed, but Exhibit 4.5 shows some of the extent of the differences. In 2003, fewer than 1 in 2 African Americans or Latino

Exhibit 4.5   Percentages of Individuals Owning Assets (Wealth) by Race: United States 2003

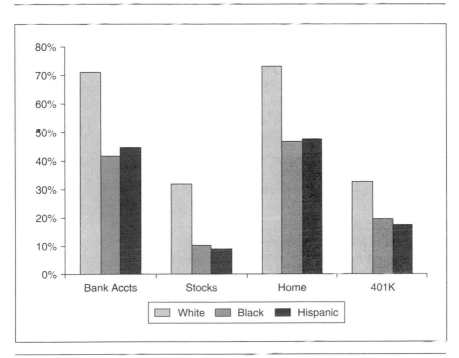

Source: Statistical Abstracts of the United States

Americans had interest-earning bank accounts or owned their own homes. Fewer than 1 in 10 owned stocks, and fewer than 1 in 20 participated in 401K retirement accounts. These statistics indicate that roughly 1 in 2 African Americans and Latino Americans are essentially propertyless—possessing virtually no financial resources to pass on to the next generation, either in the form of inheritance or in the form of enhanced resources in the home. Note also that the absence of retirement accounts inverts the relationships between parents and children, as the children of African Americans and Latinos are far more likely to funnel resources *to* their parents in later life, which in turn saps the financial resources that they might be able to send to their own children.

Wealth deprivation is especially important to consider in the contemporary economic situation because it plays a significant role in decreasing the likelihood that disadvantaged children will successfully apply to competitive college programs. Opportunities to take advanced placement exams or to engage in extracurricular activities (necessary for entry into elite college programs) are in many cases absent from rural and urban schools where African American and Latino students are concentrated. Movement to desirable school districts requires the financial wherewithal to afford down payments and hefty mortgages, often beyond the reach of disadvantaged parents (Kozol 2006). In other words, the possession of wealth (in the form of owning one's home) is directly linked to access to high-quality education.

How can the economic capital differences among racial and ethnic groups be reduced? One solution is to rework the distribution of resources so that members of the lower classes (who are disproportionately members of minority groups) have a more equitable share of the collective pool of income and wealth. This could involve, for instance, revising the minimum wage standard to enable those in low-wage jobs to earn enough to get by and to invest in their children's lives. This would also require rethinking how community resources are directed, for instance, reversing the current system so that greater proportions of social resources are directed to schools in disadvantaged (minority) neighborhoods.

Race, Ethnicity, and Human Capital

At the beginning of the movie *The Graduate*, Mr. Robinson offers Benjamin Braddock a piece of advice about how to make his fortune— "Plastics . . . enough said." The message was straightforward; all Ben needed to do was develop the skills, apply himself to that industry, and his future would be made. Such advice fits well within **human capital theory**, a perspective that focuses on the value of the skills different employees apply to their jobs. **Human capital** comprises capabilities vested in individuals, such as technical skills, education, and experience. Those individuals who have the most human capital generally are the most valued employees, earn the highest wages, and have the strongest prospects for upward mobility. Conversely, those who lack skills, or who possess skills that do not fit current opportunity structures, experience economic hardship.

One approach to documenting differences in human capital is to study achievements in formal schooling. African Americans (80%) are earning high school degrees at rates that are only modestly lower than whites (86%) and Asians (87%), a dramatic change from the past, when African Americans were far less likely to complete high school than other groups. Latinos, on the other hand, continue to lag far behind (58%), a dynamic that largely reflects the recent arrival of poorly educated immigrants from Mexico and other parts of Latin America (Waldinger 2001). But even if high school graduation rates among the various ethnic groups were at complete parity, African Americans and Latinos are far more likely to attend schools that provide inferior learning opportunities compared with those offered in the suburban schools whites and Asians are more likely to attend.

More revealing are differences in the likelihood of completing postsecondary education; these trends are illustrated in Exhibit 4.6. Although

Exhibit 4.6   Percentages Graduating College (Age 25 Years and Older) by Racial Group: United States 1960–2004

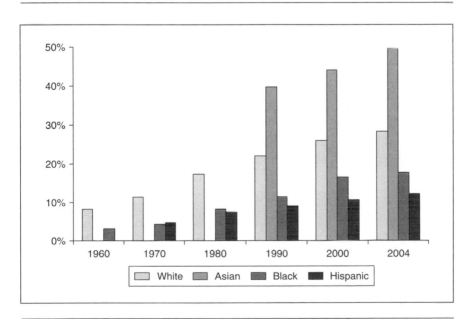

Source: Statistical Abstracts of the United States
Note: Data on Asians not available before 1990.

college completion rates have increased for all groups during the latter part of the 20th century, Asian Americans have the highest college completion rates, with whites second. African Americans and Latinos are significantly less likely to have a college degree, which puts these groups at a distinct disadvantage in the competition for the higher-skilled and higher-paying positions in the new economy.

The racial differences in educational attainment help resolve one of the major questions concerning race and work—why are so many Asian Americans (many of whom are recent immigrants) succeeding when large numbers of African Americans and Latinos are not? One explanation can be found in Asian Americans' possession of the type of human capital valued in today's economy—the skills obtained through a successful college education. They are what some sociologist call "human capital" immigrants (such as computer programmers and doctors from India), who typically adapt to the United States quickly and get jobs fairly similar to those obtained by nonimmigrants (although there is evidence that Asian Americans with degrees from outside the United States do less well than those with degrees from American universities and that "silicon ceilings" exist in high-tech firms for Indians). Latinos are much more likely to be "labor migrants," who arrive with limited education and consequently, make slower economic progress (Alba and Nee 2003; Varma 2006; Zeng and Xie 2004).

The strong effect of human capital differences on economic inequality points to the need to improve opportunities—especially for members of disadvantaged minority groups—for education and training. But this alone is not a solution to the problem of inequality because many jobs in today's economy are designed to be performed by low-skilled workers. Even if one could wave a wand and make all members of society hold PhDs, there would still be hamburgers to flip, shelves to stock, and cash registers to operate. This indicates the need for initiatives to improve the returns on work for the disproportionate shares of minority workers who remain dependent on the bad jobs in the new economy. And, as Katherine Newman has argued, this will involve not just better wages, but also various kinds of support for working people with limited resources (child care, health care, etc.) (Newman 2006).

Race, Ethnicity, and Social Capital

Human capital is about "what you know"; in contrast, **social capital** is about "who you know." Social connections to wealth and opportunity are important to the process of getting a job and moving up in the labor market. People tend to get jobs through the people they know (Granovetter 1973 1995). Imagine how much better off Benjamin Braddock would have been if Mr. Robinson had owned a plastics company!

African Americans, recent immigrants, and white Americans continue to interact in **homogamous networks**—sets of social relations with others who are similar in race and economic position. African American and immigrant networks may help them get jobs, but those jobs are likely to be low-level jobs. Those at the lower end of the class structure are less likely to have much contact with people who can help them move up occupational ladders (Fernandez and Fernandez-Mateo 2006). Moreover, among the poorest groups in the new economy (where African Americans and Latinos are overrepresented), social networks often get used, out of necessity, primarily for day-to-day survival, and can even create demands on women (who are most frequently called upon to help in times of need) that make stable employment unlikely (Dominguez and Watkins 2003; Stack 1997 [1974]).

The existence of social capital plays an important role in explaining the successes and failures of immigrant groups to prosper in the new economy. For instance, when Nicaraguan professionals migrated to Florida in the 1980s, they experienced difficulty finding good jobs, despite their high levels of education. In contrast, relatively poor and uneducated groups such as the Vietnamese who settled in Louisiana during the same period tended to do better, despite possessing weak English skills. The key difference was the strength and extent of the Vietnamese family networks that helped new members learn of job opportunities and placed them in positions in locally owned businesses (Portes and Rumbaut 2001).

Other studies reveal similar effects for other ethnic groups that have had a longer history in the United States. For example, one study compared the employment experiences of young men growing up in three ethnic neighborhoods in Brooklyn, all of whom had high risk for entering criminal careers. When young Italian American men got into trouble, male relatives helped them get into the industrial work that they themselves performed. By contrast, Latino youths tended to work various kinds of neighborhood jobs with friends and relatives (especially those involving automobiles), which led them into more organized illegal work (in "chop shops") or into auto mechanic's jobs. African American youths' networks linked them mostly to various kinds of public sector employment where connections were of less help. These youths also had fewer network resources that could keep them out of trouble (Sullivan 1989). These findings indicated that the types of ties, and the strengths of social ties, mattered in the careers of these young men.

Access to networks and social capital is also an important factor in explaining the differences between African Americans and recent Latino immigrants.

Both groups suffer a poverty of social capital, but in different ways. One problem for inner-city African American young people is that their teachers and family members have few social ties with employers and, as a consequence, have fewer doors opened for them (Kirschenmann and Neckerman 1991; Moss and Tilly 2001; Royster 2003). They also have far fewer contacts with jobholders, and, as a result, have limited information about opportunities, or mentorship on how to develop a successful career (Holzer 1991; W. J. Wilson 1997).

Interestingly, Latino immigrants in inner-city locations appear to be finding jobs, despite the shortage of available opportunities, partly because of their social ties to other Latinos who hold jobs (Mouw 2002). Actually, these networks are so effective that some workplaces have developed internal norms of referring and hiring only Latino workers. Given language barriers, these tightly knit work groups can make it difficult for employers to hire members of other ethnic groups (Waldinger and Lichter 2003). In contrast, African Americans have been less able, or less inclined, to establish this kind of ethnic enterprise so their ability to help co-ethnics in this way has been far more limited (Butler and Herring 1991). But even though inner-city Latinos are more likely to be employed than are African Americans, their network ties tend to link them to relatively low-wage jobs (Fernandez and Fernandez-Mateo 2006; Waldinger 2001; Waldinger and Der-Martirosian 2001).

Differences in social capital also make enormous differences in chances for upward mobility within organizations. One's ability to move up the ladder depends not just on merit but also on mentoring and sponsorship received from senior employees who provide resources, opportunities, and protection when the going gets rough (e.g., Jackall 1989). Members of racial and ethnic minorities are at a disadvantage in getting this kind of help partly because of their scarce representation in positions of power, which creates a dearth of the kinds of social ties that are available to well-connected white protégés (Ibarra 1993). When members of minority groups *do* have access to strong, resource-rich networks, they are much more successful in gaining access to high-level jobs and senior positions (Ibarra 1993; Zweigenhaft and Domhoff 2003). As we later discuss, despite its many problems, this is one of the primary justifications for the continued application of affirmative action policy.

Race, Ethnicity, and Cultural Capital

The best jobs in the new economy require not only technical skills, but also the ability to work effectively with people. This requires knowing when and how to discuss issues, and how to comport oneself appropriately—especially when interacting with others who can open opportunities. The focus on **cultural capital**—general cultural orientations and dispositions that can be used for gain—shows that it is not simply the ability to work, but also the ability to "fit in" that influences success (Bourdieu 1986). Acquiring cultural capital requires more than simply earning a college degree; it necessitates the ability to harmonize interaction, create friendships, know the right conversational skills, operate within taken-for-granted behavioral norms, and master other "intangibles" that are not part of formal job descriptions.

Culture operates like a toolkit, enabling individuals to navigate the worlds in which they live (Swidler 1986). But depending on the social world, the tools in that kit will vary. Members of racial and ethnic minority groups are more likely to come from families at the bottom of the economic ladder, so many have not been initiated into middle-class culture by either their families or the schools they attended. Consider, for instance, what it is like to be a young man growing up in a drug-infested inner-city neighborhood. In this world, the code of the street is to gain and maintain "respect," and success in the neighborhood requires toughness and the ability to exert masculine dominance. But though these cultural skills facilitate survival in the neighborhood, they prove counterproductive to presenting oneself in ways that would be appropriate to work in downtown offices. When these young men venture into this world, they feel awkward and inadequate, aware that they do not fit in and that their language and attire is too "street." They also find it difficult to take orders or tolerate indignities, especially those they perceive as coming from female supervisors (Anderson 1999; Bourgois 1995). Nor is this simply a matter of their own attitudes; employers, too, look for the "right" cultural capital in prospective employees. "Unconventional" forms of dress or appearance (braids, dreadlocks, beaded hair, etc.), nonstandard English, and other characteristics of inner-city African American life are often looked down upon by employers, who see these as markers of unreliability and an inability to fit in (Moss and Tilly 2001).

Cultural capital is not learned in the same way one might learn computer programming. It comes from years of socialization that form a **habitus**—the habitual ways of interacting with and understanding one's world (Bourdieu 1984). This requires more than memorizing whether forks go on the left or the right. It requires the development of a natural ease in the situations that matter and the ability to belong with those who count. And it involves developing perspectives about what to expect for oneself and what to expect from others. One study of successful African Americans found that they stressed the need to learn to fit into the social world of more affluent whites. These African Americans argued that their experiences in private schools had taught them to be confident and comfortable in a social environment very different from the one into which they were born (Zweigenhaft and Domhoff 2003).

Analyses of lower-class minority youths indicate that they tend to be very different, in terms of cultural capital, from these successful African Americans and the successful whites with whom they now interact. Classic studies of lower-class young men indicate that they often lack a clear sense what types of jobs are available, may have unrealistic aspirations, and often develop a habitus of self-blame when things don't work out as they had hoped (Liebow 1967; MacLeod 1995). A recent study of young African American men living in the city of Chicago found that they held fairly conventional (if often unrealistic) goals in a community where it was unlikely that they would find good work. They wanted "good jobs" that brought respect; they talked about city jobs, starting small businesses, and, notably, getting blue-collar jobs in factories, even though few such jobs remained in the communities in which they lived. However, when asked what they *expected* to be doing in a few years' time, they had no clear sense of the future or of how their aspirations would pan out. They possessed only a sketchy knowledge of efforts that might benefit them in obtaining work in the contemporary economy. In sum, they thought they could succeed but lacked the cultural capital to know how (Young 1999, 2003).

The importance of cultural capital in an economy where most work involves social interaction should not be underestimated. But how does one teach culture? Part of the problem is the extent of economic segregation, which has increased since 1970. As long as those from poor backgrounds interact largely with others in similar situations, and as long as community environments promote different cultural codes and values, this will remain an enduring problem.

Geographic Distribution of Race and Work Opportunity

Understanding the interaction of race and work requires considering residential patterns, the geographic distribution of work, and the prospect that individuals will be able to find good jobs that match the skills they possess. Although proximity does not guarantee that one will get a job, the chances of finding a good job will be better if the jobs are located near one's home and community, and if one resides next to other people who have jobs. The so-called spatial mismatch hypothesis, that is, the idea that people often have trouble getting jobs if they are located where employment is scarce, has been shown by many studies to help explain why minority workers (who are often spatially concentrated) have limited access to good employment (Fernandez and Su 2004).

One of the most striking examples of how geography has affected the economic experiences of an ethnic group can be found in the case of Native Americans. The reservation system, established in the 19th century, virtually ensured that indigenous communities, established in remote rural areas, would be at the margins of the economy. The lack of work is one of the primary causes for this ethnic group's escalated risk for suicide, alcoholism, and out-of-wedlock births (Erikson 1994; Snipp and Summers 1992).

William Julius Wilson (1987, 1997) demonstrated comparable effects on the experiences of African Americans in the Northeast and Midwest. In the middle of the 20th century, when the industrial economies of cities like Detroit, Cleveland and Buffalo were booming, African American men found work in factories, and their families created stable communities. However, from the 1970s to the 1990s, those factory jobs disappeared, followed by the middle class, which left many neighborhoods in central cities populated by relatively low-income minority populations (Barlett and Steele 1992; Bensman and Lynch 1987; Bluestone and Harrison 1982). The same types of economic and demographic shifts also occurred across much of rural America (Duncan 1992). In many of these places, a "hyperghettoization" of concentrated poverty resulted in increased crime, drug use, and out-of-wedlock births, which in turn undermined the prospects of these communities' nurturing new workers or attracting new employers (Sampson, Morenoff, and Gannon-Rowley 2002; Sampson, Raudenbush, and Earls 1997).

Part of the challenge for the new economy is shifting work opportunities to the communities that are currently work-poor and facilitating the movement of families to locations where jobs are plentiful (Newman 2006). The Gautreux Program in Chicago illustrates what can happen when people are moved to employment-rich communities. It was designed to move some residents of public housing in the city of Chicago into private sector housing, and thereby reduce the impact of concentrated poverty on the lives of the disadvantaged (W. J. Wilson 1997). Half of the participants were randomly selected to move into suburban housing, and the other half were moved into housing within the city of Chicago. Those who had moved to the suburbs were significantly more likely to become employed. There were various reasons for this—including the transplants' sense that the neighborhood was safer (so leaving their kids alone was less of a problem) and that *not* working was socially unacceptable. But perhaps the most powerful reason for the change in work outcomes was that the suburbs offered more opportunity to work and that the African American transplants had little trouble finding jobs for which they were qualified (something that could not be said for those who remained in the city). Other studies have shown that moving to suburban locations does not, by itself, always improve the employment prospects of African Americans (Fernandez and Su 2004). But the Gautreux experience provides strong evidence that it *can*.

If many African Americans are disadvantaged by living in declining inner-city neighborhoods where employment opportunities are scarce, what about immigrants, who also are likely to settle in urban areas? Here, the nature of ethnic communities produces complex, perhaps contradictory effects. In some cases, ethnic communities provide opportunities for migrants to find work in businesses owned by (or at least dominated by) others from the same ethnic group. This can provide immigrants with better jobs than they would have found outside the ethnic economy (Wilson and Portes 1980). For example, Cubans who migrated to the Miami area established a significant number of businesses, most of which were small, but which employed many of the Cuban Americans who settled there. Those Cuban Americans who worked in the Cuban "enclave" were not all trapped in low-paying jobs. The ethnic enclave provided significant opportunities for many to learn new skills, including business skills, and open their own businesses, serving as an avenue to economic success for at least some immigrants.

In other cases, working for or with co-ethnics has been a real disadvantage in the longer term, trapping the immigrant in a subculture and limiting opportunities to move into the economic mainstream (Alba and Nee 2003; Bonacich 1972; Waldinger and Der-Martirosian 2001). For example, Asian immigrants in the Los Angeles area benefited from their ethnic enclave, at least at first, because it provided them with employment and other benefits when their language and cultural skills were still poor. However, the kinds of jobs they found in the ethnic economy typically provided low wages. If they remained within the ethnic enclave, they became trapped in low-wage undesirable employment, unless they were among the few who successfully established businesses. Fortunately, Asian immigrants have been largely successful, over time, in moving out into the larger economy as they become more acculturated. Their children, in turn, are even more likely to enjoy upward mobility and improved wages as a result (Nee, Sanders, and Sernau 1994).

People work in jobs, but live in communities. Ethnicity plays a role in determining where people choose to live, and ethnic enclaves create unique dynamics in shaping what types of goods and services are desired, which in turn influences the types of work opportunities that will be created. However, the movement of jobs into and out of areas where ethnic minorities are concentrated can have a profound impact on current and future generations of minority workers. As W. J. Wilson (1997) found, when work disappears, it sets the stage for a wide variety of social problems to emerge. The solution, from a work opportunity perspective, is to redistribute work throughout the various communities, and to consider not just the racial characteristics of workers, but also the patterns of racial and ethnic concentration in neighborhoods and cities.

Racial Discrimination

The creation of a truly inclusive society requires that there be opportunities to work, that individuals have the resources to develop the skills to gain employment, and that, once employed, they are treated in a fair and equitable manner. In this section, we consider the extent to which members of racial and ethnic minorities experience employment discrimination.

Prejudice and Discrimination

Officially, employment discrimination is illegal in the United States. Civil Rights–era laws that stipulate that employers should not discriminate against potential or current employees on the basis of their race (or gender) have resulted in a genuine decline in overt forms of racial exclusion (e.g., the exclusion of African Americans from apprenticeship programs or professional training programs). But the research evidence indicates that race has not been entirely eliminated from employers' decisions about whom to hire and promote.

Employers are surprisingly willing to admit to generalizing about members of racial groups and to express a preference for (or reluctance to) hire members of different racial groups as employees. Though not all feel this way, many employers still have strongly negative—and often hostile—opinions of African American employees, particularly African American men, many of whom employers regard as unreliable and difficult employees. The harshest appraisals are directed at African Americans from the inner city—that is, those who live in "bad" neighborhoods, who attended urban public schools, and who speak "Black English" and dress distinctively. Employers associate inner-city residents with a variety of negative behaviors (drug use, crime, poor education, etc.) and, therefore, express reluctance to hire people from those neighborhoods. Employers respond more negatively to job applicants with inner-city addresses or with African American–sounding names, like Jamal or Lakisha (Bertrand and Mullainathan 2004; Fernandez and Su 2004; Kirschenmann and Neckerman 1991).

Negative views of African American employees seem to focus particularly on African American men. However, a somewhat different (but also negative) perception of African American women has consequences for their chances of finding good jobs. Employers do not think of African American women as lazy or "difficult" in the way that employers tend to stereotype African American men (Browne and Kennelly 1999; Kennelly 1999). Actually, employers are more likely to perceive African American women positively, in some respects, emphasizing their role as breadwinners for their families and their role as parents. Unfortunately, this is combined with a tendency to stereotype African American women as single mothers and to project onto them problems that employers imagine go along with this status. For example, employers express concerns about tardiness and frequent absences from work in connection with African American women, believing that single parenthood will create complex work–family conflicts that cannot be resolved. Although many African American women are *not* single mothers, and although married women *and* men experience problems associated with work–family conflict, employers continue to perceive African American women as single mothers and to have concerns about their value as employees as a result.

For many entry-level jobs in particular, employers place a great deal of emphasis on applicants' "soft skills," including communication and people skills as well as qualities such as "attitude" and "personality" (Moss and Tilly 2001). Employers generally have a negative view of African Americans' soft skills, believing that they are less reliable, more likely to "have an attitude," and experience difficulties interacting with members of other racial groups. As a result, employers often are reluctant to hire African Americans for jobs that demand soft skills. In some cases, employers try to exclude African Americans altogether; in others, employers try to "match" the characteristics of the employee to the racial characteristics of the workplace or clientele (e.g., by choosing to hire African Americans for settings in which they work with or serve other African Americans). The result is different, and diminished, employment opportunities for African Americans.

Recent increases in Latino immigration (both legal and illegal) have led to questions about whether employers prefer Latino employees to African Americans. There has been some suspicion that employers are choosing to hire Latinos for jobs once held by African Americans, such that Latinos are benefiting economically at the expense of African Americans. But the research evidence suggests that the reality is more complex.

Employers perceive Latino workers as hardworking and deferential (Waldinger and Lichter 2003). Given the prevalence of negative views of African Americans as workers, it would not be surprising to find that employers prefer Latinos for jobs where skill demands are low and the main requirements are hard work and subservience. That these preferences exist, however, does not mean that Latino workers are displacing African American workers, as some suspect. To begin with, employers' preferences vary according to the type of position they are trying to fill (Waldinger and Lichter 2003). When skills or human capital are not an issue, the preference for Latinos takes precedence. However, for other jobs, Latinos may be less desired. When jobs require English-language skills, employers may actually *prefer* African Americans. Jobs in health care, for example, require the ability to interact with highly educated English speakers (doctors and sometimes patients), to read instructions on technically sophisticated equipment and materials, and to understand safety directions.

Moreover, even where Latinos are engaged in jobs that once were performed by African Americans (e.g., in light manufacturing), one cannot assume displacement has occurred. What often is happening is that African Americans have voluntarily abandoned these jobs in preference for better positions that have become available (Lim 2001). The willingness of some immigrants to accept low wages reduces the pressure on employers to pay more (which would make these jobs more attractive to other ethnic groups). But the impression that low-wage immigrants "push out" African Americans (who then become unemployed) may be an oversimplification.

Eliminating stereotypical views of racial and ethnic groups is not something that the passage of a law or the implementation of a sensitivity program alone can achieve. However, developing and enforcing effective policies for combating the discrimination that results from prejudice *does* increase opportunity (Stainback, Robinson, and Tomaskovic-Devey 2005). And the belief that people cannot do something tends to weaken when they are able to demonstrate that they can.

Racialized Jobs

Another facet of the effect of racial attitudes on patterns of employment is the existence of **racialized jobs**—jobs that are typecast to be held by members of specific racial or ethnic groups. One familiar example of this phenomenon can be seen in personal service jobs—that is, jobs like those held by maids who perform work within other people's households. There is a long tradition in the United States of expecting African American and Latina women to perform domestic labor for others. Although racial attitudes have changed, race still affects maids' work experiences. African American and Latina maids are much more likely than are whites to be asked by employers to perform personal services, to engage in stigmatizing tasks (changing bedclothes, washing underwear, scrubbing on one's hands and knees), and to experience a variety of intentional or unintentional slights at work (Wrigley 1999).

"Racialized" jobs and work roles can also be seen in higher-level occupations. The case of the growing African American middle class is illustrative. In the past, the African American middle class tended to be confined to a racial niche. African American professionals, business owners, and the like existed, but they were largely concentrated in African American neighborhoods, providing services to African American customers and clients.

In the last few decades, middle-class African Americans have been finding their way into predominately white organizations, especially in the public sector. However, they also are being funneled into smaller, peripheral firms (Vallas 2003b), have less authority (Elliott and Smith 2001), earn less even when they are in positions of authority (Smith 1997), and are more likely to supervise African Americans than whites (Elliott and Smith 2001). Evidently, being an African American professional or manager is not exactly the same experience as being a white one.

Many African Americans who enter the corporate world encounter a contradictory reality (Collins 1997). On the one hand, a desire to integrate workplaces, combined with affirmative action programs, makes minorities highly desired, and companies often actively seek these types of employees. However, even though the obstacles to entering the corporate world may have been reduced, what happens afterward is different. Many African Americans find that, once they have been hired, their progress up the corporate ladder is slow. It seems to take longer to get promoted, and they suspect that they have to meet a higher standard than do their white counterparts with similar kinds of education and experience (G. Wilson 1997). They also find themselves "steered" toward dead-end jobs in areas such as community relations and diversity offices, which offer attractive starting pay and a chance to "give back" to the community, but provide few avenues for career growth (Collins 1997).

Race, Ethnicity, and Work: Social Policy

The persistence of racial and ethnic inequality at work, and the tensions and conflicts surrounding its resolution, raise a host of policy issues. Earlier, we examined the dominant theories regarding why some racial and ethnic groups lag behind and why others are advancing in the new economy; for each theory, we highlighted the implications for policy reform. Here we consider two pressing policy debates—the use of affirmative action programs and how immigration affects work opportunity in the United States.

Affirmative Action

Affirmative action is the most visible effort in the last 40 years to confront racial (and gender) inequality in the workplace. In 1968, the federal government began requiring firms with significant government contracts ($50,000 or more) and at least 50 employees to produce annual affirmative action reports and to demonstrate a commitment to hiring minority workers (including women). This became the basis of the programs we now know as affirmative action, which have been subsequently adopted by many companies, as well as by colleges and universities in their selection and recruitment of students (Stainback et al. 2005).

How does affirmative action work? In most cases, the policy simply requires that employers make a "good faith effort" to recruit and retain employees from underrepresented groups—specific minorities and women. This means that employers must advertise vacancies in ways visible to minority candidates, explain why they did not hire minority candidates if qualified candidates applied for the positions, and produce regular affirmative action reports documenting their efforts.

Opponents of affirmative action label it as a quota system and conclude that it leads to the hiring of unqualified individuals. Opponents express concerns that affirmative action leads to "reverse discrimination," wherein minority candidates are favored over whites in hiring and promotion decisions. Individuals not favored by the program commonly conclude that difficulties experienced in securing work result from opportunities being unfairly directed to minorities and women. Even some minority group members express concern that they are stigmatized by assumptions that they have been hired only because of their skin color (Steele 1990). National opinion polls, such as the General Social Survey, reveal that most whites do not favor affirmative action and that the African American community is divided on the issue.

On the other hand, proponents argue that affirmative action does not operate on the basis of quotas because specific positions are not set aside for minority candidates. Proponents also argue that affirmative action does not undermine the appointment of qualified individuals. To be hired under affirmative action programs, all candidates must meet base level qualifications. Once those criteria are met, race and gender need to be considered for two reasons. First, it is in the interests of society to distribute opportunities more equally, and this system (despite its flaws) is needed to combat discrimination and other factors that limit opportunity. Second, creating a diverse workforce or student body enhances organizational functioning in a multicultural society. National survey data indicate that although Americans generally oppose policies called "affirmative action," they are actually sympathetic to efforts to seek out and retain minority employees (Bowen and Bok 1998; Reskin 1998).

To date, the courts have upheld the constitutionality of affirmative action. But they have put constraints on how it is practiced by ruling that seniority rules take precedence over affirmative action (*Firefighters v. Stotts*, 467 US 561 [1984]), by shifting the burden of proof in claims of discrimination to the person making the complaint (which makes it much more difficult to prove) (*Ward's Cove Packing Co. v. Antonio*, 490 US 642 [1989]), and by making it easier for employees to claim that affirmative action resulted in reverse discrimination (*Martin v. Wilks*, 490 US 755 [1989]). The recent Supreme Court decision regarding the University of Michigan's affirmative action policies did not abolish the program altogether, but ruled that the school could not continue to use a rating system for candidates in which points were awarded simply for minority status.

One of the most important questions today concerns the effectiveness of affirmative action programs—are they opening doors that would otherwise be closed? Strong evidence indicates that affirmative action policies have helped African American and female workers—especially those from the middle class—by pressuring major employers to alter their hiring practices (Reskin 1998). The most visible effects of affirmative action have been in the public sector, where minority employment has increased significantly, partly because of the greater effectiveness of enforcement. However, the effects have been less noticeable in recent years, as the political desire to enforce them has eroded (Stainback et al. 2005).

However, there are limits to the success of affirmative action. One concern is that the programs primarily help African Americans, Latinos, and women who possess college degrees because they force employers to consider them for higher skilled jobs. In contrast, affirmative action has been of only limited help to people at the bottom of the class structure, those who lack the qualifications for desirable employment. Additionally, although affirmative action opens some doors into workplaces, it does not combat the glass ceilings, "racialized jobs," and dead ends that minority and female employees often encounter after they enter large organizations and attempt to move up the ladder (Collins 1997).

Immigration Policy

A second policy issue that has dominated the recent headlines concerns what to do about immigration. Rising numbers of immigrants have led to concern about their economic impact and to calls for a return to a more restrictive approach. Much of the concern focuses on illegal immigration, but many Americans seem worried, especially in a period of stagnating wages and rising insecurity, that immigration *in general* has gotten out of hand.

A primary source of hostility to immigration is the belief that immigrants compete with American citizens for jobs. Economists and others have argued long and hard about the economic effects of immigration. Some conclude that the effects are largely negative, at least for American workers. They argue that an increase in immigration tends to result in lower wages because immigrants are willing to work "for less" and are vulnerable to unscrupulous employers. In addition, the growth in the supply of workers reduces the pressure on employers to raise wages to attract employees. Others argue that the situation is more complex. Areas with high rates of immigration (including *illegal* immigration) are not atypical in their wage structures (Broder 2006; Lowenstein 2006).

Analyses of local labor markets indicate that the effects of immigration on wages in particular areas tend to be slight (Sassen 1995). Some have argued that the real effect is at the national level; they argue that if immigration had not occurred, the wages earned by low-skill workers in particular would have been significantly higher (Borjas, Freeman, Katz, DiNardo, and Abowd 1997). However, these studies are inconclusive, and some economists conclude that the safest assessment is that immigration has had only a modest negative impact on the wage growth of low-skilled individuals in the United States (and a somewhat positive effect on others' wages) (Card 2001; Saiz 2003).

Other critics suggest that the problem with immigration is that it results in higher unemployment. Immigration increases the labor supply, creating more competition for jobs, and may encourage employers to replace Americans with less expensive immigrant labor. The evidence in support of this argument is *less* persuasive than on the issue of wages. First, it is difficult to disentangle the economic effects of immigration from the effects of other economic changes; for example, if unemployment rises in an area that has experienced both immigration *and* plant closings, which of the two is responsible? In addition, immigration is often associated with economic growth, so even if workers are displaced by new migrants, an overall expansion of employment may create new job opportunities for those who have been displaced. A classic study of a sudden increase in immigration in a local labor market (the Mariel immigration of Cubans to South Florida in the late 1980s) found that the labor market adapted to the influx of immigrants, absorbed them successfully, and that unemployment was unaffected (Card 1990). Workers, particularly immigrants themselves, also move when employment prospects are limited in a particular area, so an influx of immigrants does not typically lead to a growth in unemployment (Saiz 2003).

If the effects of legal immigration are complex, what about *illegal* immigration? The United States, like many countries, has been experiencing a high level of illegal immigration, primarily (although not exclusively) from Mexico and other Latin American countries. From 1992 to 2005, the population of illegal immigrants in the United States rose 185% to just over 11.1 million (Broder 2006). This large undocumented population has generated much controversy, particularly in the states where it is concentrated, and has led to calls for new efforts to prevent further immigration of this kind.

The growth of immigration from Latin America has been fueled by a variety of causes. The Bracero program introduced many Mexican families and communities to the American labor market as temporary guest workers, but when the program ended in 1964, the immigration continued in different forms. Changes in immigration law in the 1960s made it easier for those in the United States to sponsor relatives as immigrants. And changes in the economy of Mexico (particularly in agriculture, where traditional small farmers confronted more-efficient, better-capitalized farms producing for global markets) pushed more Mexicans to leave their communities or to send family members away in search of income (Alba and Nee 2003).

Changes in U.S. immigration policy have gradually made it more difficult for relatively unskilled Latin American migrants to enter the United States legally. The passage of the 1986 Immigration Reform and Control Act attempted to restrict the flow of illegal immigrants through a combination of increased border patrols and stricter sanctions on employers, who were made subject to penalties for hiring undocumented workers. However, the act was largely ineffective, partly because the enforcement of penalties against employers remain almost nonexistent. Additionally, employers are increasingly hiring undocumented workers through subcontractors, thereby making it much harder to detect and providing a shield against incrimination (Alba and Nee 2003; Massey and Bartley 2005). Since then, most efforts to control illegal immigration have focused on stricter control of the border, an effort that appears not to be working and that has forced people to attempt to enter the country through more dangerous, sparsely populated arid environments. There is little evidence that increased apprehension of illegal immigrants has discouraged these individuals from making an initial attempt, or from trying again (Cornelius 2005; Massey 2006). Moreover, focusing on interdiction at the border ignores the fact that many undocumented workers enter the country legally, on short-term visas, and simply stay on once their visas have expired. Building walls between Mexico and the United States does little to affect this aspect of the problem.

The most urgent question regarding illegal immigration concerns its effects on the United States, its economy, and its workers. As Exhibit 4.7 shows, illegal immigrants compose sizable proportions of the labor force in jobs that require low-skilled work, especially in farming, cleaning, construction, and food preparation, where between 5% and 25% of workers are illegal immigrants.

Exhibit 4.7   Percentages of Occupations Filled by Illegal Immigrants: United States 2004

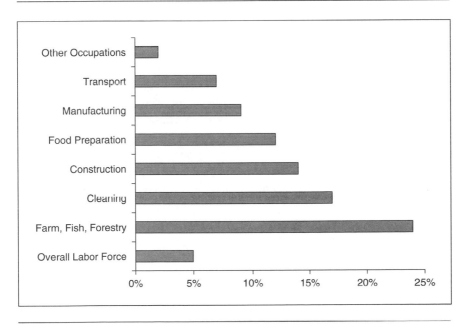

Source: Broder (2006)

Illegal immigration is believed to allow large numbers of undocumented aliens to enter the United States, drain public resources, take away jobs from legal residents of the United States, and exert a downward pressure on wages and working conditions (Buchanan 2006). Others object to these assumptions, noting that high levels of immigration (legal or illegal) are often linked to economic prosperity. For example, undocumented workers enable the production of affordable consumer goods, partly because they accept terms of employment deemed unacceptable by legal residents. It also bears noting that illegal immigrants often pay taxes but do not use the public services to which they are entitled (either out of fear or because they lack required documents). When illegal immigration drops, furthermore, American workers do not rush to fill the jobs that theoretically become available; on the contrary, employers often wind up pressuring politicians to loosen immigration laws so that they can obtain needed labor. And immigration proponents argue that immigrants, legal or illegal, benefit the American economy because they are hardworking, motivated, and enterprising (Broder 2006; Lowenstein 2006).

It is difficult to assess the extent to which illegal immigration depresses wages because evidence is contradictory on this point. It is possible that employers do not raise wages because the supply of cheap illegal migrants exists. Alternatively, it is possible that these are simply low-wage jobs that would go unfilled if migrants stopped coming. Some support for the first argument can be found in the reality that it would not be all that difficult, economically, to raise wages in sectors where illegal immigrants work. For example, wages represent a tiny percentage of the cost of the fruits and vegetables that Americans consume. Even raising wages significantly in that sector would result in a small (2% or 3%) increase in the cost of produce; thus, the argument that employers *can't* raise wages rings somewhat hollow (Broder 2006).

What should be done about illegal immigration? Most of the discussion has focused on stricter border enforcement, an approach that has, thus far, proven to be ineffective. Arresting and deporting large numbers of illegal immigrants also has been discussed, but the evidence suggests that many deportees simply return. Moreover, many illegal immigrants have children who were born in the United States, creating a moral dilemma for those seeking to deport the parents. The alternative is to focus on employers by strictly enforcing sanctions against those who hire illegal immigrants. This would have the effect of reducing demand for illegal immigrants, which, at the moment, remains high (Cornelius 2005). Additionally, as we discussed previously in this book, policy needs to respond not only to "pull factors" (what is attracting these workers to the work in the United States), but also the "push factors" (labor problems in the developing world and how poor job prospects at home compel workers in these societies to seek work elsewhere). A domestic work policy, especially as it relates to the issue of illegal immigration, would have to be coupled with an international work policy that ensures opportunity for workers irrespective of their nationalities (Kochan 2005).

Conclusion

Race, ethnicity, and work are commonly thought of as issues of the past, especially by those who are not directly or indirectly discriminated against. In this chapter, we showed that the opportunities to work, and to succeed at work, remain strongly tied to issues of race and ethnicity, both because old problems persist and because new ones have emerged. They influence opportunities to obtain resources, to develop the skills needed to get good jobs, to tap into social networks that open doors, and to fit into work environments dominated by white (and male) power holders. Race and ethnicity also influence the opportunity to find jobs, which are often far removed from ethnic communities.

To shape a work policy for the new economy—one that is truly inclusive of all groups—race and ethnicity need to be included as important considerations. Part of the solution will necessarily involve creating greater equality of opportunity and directing more resources to minority groups that suffer greater poverties (of finances, skills, and social connections). This will require increased allocation of funds to underprivileged families and communities. The persistence of racial and ethnic preferences in hiring and the continued application of racial and ethnic stereotypes point to the need for further development of multicultural sensitivities. This will be challenging to accomplish as long as geographies divide the population into polarized ethnic enclaves. And efforts need to be made to promote the creation of good jobs in places where employment has largely disappeared. But perhaps some of the most effective initiatives to reduce racial inequalities will involve creating a situation where those at the bottom of the economy can be adequately rewarded in this affluent society. This can be done by increasing the resources that come from jobs (e.g., by raising the minimum wage to a living wage) and by providing collective resources such as health insurance to all, irrespective of their racial identities. This would help poor people of *all* racial and ethnic backgrounds, and would do much to promote racial equality as well.

There has been remarkable progress in advancing racial inclusion at work, but old interracial tensions remain and new ones are emerging. Far from having eliminated the problem, the new economy remains divided along racial and ethnic lines, and there are few indications that the long history of conflict over these issues is drawing to an end.

Notes

1. The U.S. Census treats "Hispanic" as a broad ethnic category that includes many different groups (Mexican Americans, Cuban Americans, Puerto Ricans, etc.). Some have objected to this term, both because it is "artificial" (clustering together groups who think of themselves as different) and because it includes groups who don't speak Spanish (for example, many Guatemalans or Peruvians speak native languages rather than Spanish). The term "Latino" has been suggested as an alternative, although there is no consensus about which term is preferable. We have used the term "Latino" primarily to emphasize the issue of ethnic diversities. However, in the tables, we have used the term "Hispanic," as it is used in the government sources from which the data originate.

2. Because costs of living change over time, we converted these income figures to "constant dollars"—the values of these earnings are adjusted for the effects of inflation.

3. It should be added that these data consider only people who have incomes. Because African Americans are much less likely to be employed than either whites or Latinos, these data probably understate the gap in income between the average African American and the average member of other groups in American society.

5

Whose Jobs Are Secure?

J obs and workers are changing in the new economy and so are the prospects that workers will be able to keep these jobs from week to week, month to month, or year to year (Smith 2002; Smith and Rubin 1997; Valletta 2000). Today, insecurity has spread throughout the economy and affects a widening spectrum of workers. These include people like Jamal, the fast-food worker, who labor in traditionally unstable jobs. But insecurity has also crept into the lives of skilled professional workers like Eileen, who labor in the growth sectors of the economy. And Chi-Ying's opportunities (and insecurities) in a developing society are intricately connected to the declining opportunities for manufacturing workers like Dan, the former autoworker (Castells 2000; Heckscher 1996; Mattsson 2003; Milberg 2004; Neumark 2000). The technological and organizational transformations that created the new economy have also intensified various kinds of risk. As a result, working families face increasing levels of stress in keeping careers intact (Beck 2000; Hacker 2006; Sweet, Moen, and Meiksins 2007).

In this chapter, we focus on the issue of job insecurity and how the prospect or experience of job loss shapes workers' lives and experiences. To do this, we first consider how job and career threats were intertwined in the old economy, and how the approaches to managing these perils in the United States differed from those adopted in many European countries. We then consider workers' exposure to risk in the new economy and gauge the extent to which job insecurity has grown. As we will show, the spread of job insecurity has occurred because of the decline of older, more secure types of jobs, but also because of changing strategies for organizing work and the changing composition of the labor force. The result is that more families are grappling with the economic and emotional turmoil that results from job loss, and they do so in a society ill-equipped to meet their needs.

Risk and Work: Historical and Comparative Views

One critical question in the sociology of work concerns who bears the burden of risk. For instance, if production orders fall, is it the employer's responsibility to continue to provide paychecks to workers? If a new technology can be developed to replace workers, is it the employer's responsibility to find obsolete workers new jobs? If an applicant is willing to work for less than an existing employee, does that employee have any right to keep that job? And in an economy built on wage labor, should citizens receive support if no jobs are available? In the wake of the Industrial Revolution in both Europe and America, the answers to all of these questions would have been "no." Employers were expected to provide jobs, but at their discretion and according to terms and conditions of their own choosing, as well as in respect to what the labor market would bear. Government was not charged with the responsibility of providing assistance to those who lacked jobs or to those who could not work.

In the early phases of industrialization, workers labored without any of the protections that are present today. To secure work, workers commonly had to bribe foremen for jobs, and they could expect discrimination on the basis of their race or ethnicity. If they refused to give their foremen kickbacks, or if they caused any trouble on the shop floor, they could expect to be vindictively dismissed. And if they were cast out onto the street, there were virtually no government-sponsored social supports to help them find work or survive extended joblessness (Engels 1936 [1845]; Jacoby 1985; Katz 1996; Thompson 1963; Trattner 1999). Even as the problems of an emerging industrial economy became increasingly apparent, U.S. politicians at the national level remained reluctant to involve themselves in workplace affairs or tackle the issue of unemployment (Lipset 1996; Skocpol 1992). In part, this inaction was the result of individualistic values deeply ingrained in American culture, beliefs that individuals are primarily responsible for their own fates and that government regulation impinges on personal liberty (Bellah, Madsen, Sullivan, Swidler, and Tipton 1985; Galbraith 1976; Tocqueville 1969 [1836]). Political inaction also reflected the control early industrialists had over political processes and their ability to discourage legislation to regulate the terms under which work was performed. In sum, in the early phases of industrialization, risk in America was primarily shouldered by individual workers.

In the United States, a dearth of federal government programs to protect workers created opportunities for entrepreneurs to offer risk management services in the form of **private insurance.** Although some forms of insurance had been used by industrialists (particularly those in the shipping industries), it had not been something available to workers. But by the early 20th century, workers could pay premiums to purchase health and long-term disability insurance (Jacoby 2001). During the 20th century, these insurance policies (sometimes purchased by individuals, sometimes offered as a job benefit) remained one of the central means of managing the risks associated with job loss or the prospects of disability. Today, this private insurance approach to handling risk has come to haunt the American worker because insurance companies are powerful political forces that block government-sponsored health and job protection programs. And private insurance is becoming less accessible because rapidly rising premiums put it beyond the reach of low-wage workers and many employers.

During the 19th and 20th centuries, workers also developed collective ethnic and class-based strategies for mitigating the risks of personal hardship. Starting in the 1840s, a variety of fraternal organizations, such as the Masons and the Odd Fellows, emerged. These urban social clubs formed along ethnic lines and provided member families with small death benefits and, in some instances, health insurance. This **mutual assistance** approach to managing risk, whereby collective resources were pooled to provide support in times of hardship, was later adopted by the trade unions that formed in the second half of the 19th century. But unlike the ethnic affiliations of the early fraternities, trade unions distributed mutual assistance to members of occupational groups, thus providing skilled workers with resources that could help them survive unemployment and disability (Jacoby 2001).

As organized labor expanded its power and membership during the 20th century, collective bargaining agreements provided unionized workers with some protection from the risk of job loss. Unions fought for limits on arbitrary dismissal with victories including the implementation of seniority rights (so that veteran employees could not be laid off as easily) and grievance procedures (that discouraged vindictive attacks on individual workers) (Edwards 1979). Unions helped large portions of the workforce gain access to jobs that were well paid and stable. But union initiatives in the management of risk focused primarily on advancing their members' interests. Although some nonunion employers did emulate unionized firms, many nonunionized workers (who were disproportionately women and minorities) were left lacking social supports and job protections (Jacoby 2001; Lichtenstein 2002).

In the United States, access to job security and benefits is heavily dependent on what kind of job workers have and for whom they work. Their ability to "make ends meet" requires holding a job, and, if that job is lost, finding a replacement job quickly. In contrast, the approach adopted in most European countries was to address the problem of risk by introducing **entitlements**—rights and resources available to all citizens independent of attachment to the labor force. These transfers from the government to individuals take a variety of forms, including assistance for unemployment, disability, sickness, old age, housing, family leave, and child care.

Exhibit 5.1 reveals that only 8% of the U.S. gross domestic product is directed to assistance, roughly half the percentage budgeted in most Western

Exhibit 5.1   Government Cash Transfers as Percentages of GDP: International Comparisons

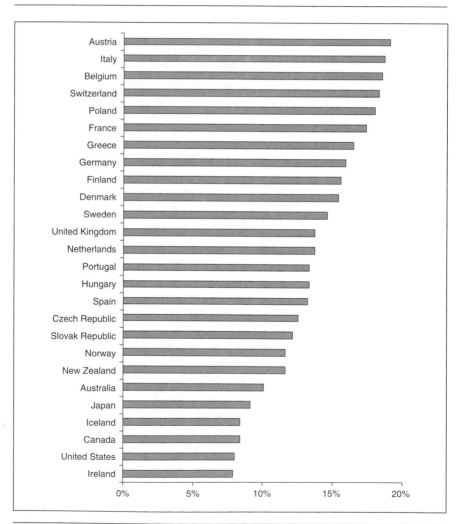

Source: Organisation for Economic Co-Operation and Development (2006)

European countries. As a result, workers in Europe have access to a greater quantity of resources that can be used if they lose jobs. In most of these countries, all citizens—not just those employed in good jobs—have access to health insurance. And in many European countries, employers have to justify displacing workers and cannot do so with the impunity extended to U.S. employers. As a result, workers in America today assume far greater amounts of individual risk in comparison with workers in Europe.

American workers do have some protections (see Appendix A for a regulatory timeline). The U.S. government's first major advances in the management of risk came as a result of the large-scale job losses that occurred during the Great Depression. In 1935, the **Social Security Act** introduced supports for those who cannot work, including aid to the elderly, the disabled, and some children. The **Fair Labor Standards Act**, implemented in 1938, introduced many of today's most important workplace regulations, including short-term unemployment insurance and a national minimum wage. Legislative acts such as these provided vital assistance to American workers. But again, these protections are much less extensive than those enjoyed by workers in other industrialized countries.

The primary buffer against job loss for American workers is **unemployment insurance**, a program operated through federal and state government partnerships (U.S. Department of Labor 2007). The goal of unemployment insurance is to provide short-term assistance so that workers have the financial resources to seek suitable replacement jobs (Leana and Feldman 1995). Because the program is administered by individual states, eligibility and compensation vary within the nation. Most states provide partial wage replacement (most commonly 50% of wages—with a cap for high income earners) for a maximum of 26 weeks. Exhibit 5.2 shows that this compensation is meager in comparison with what is available to workers in most other developed countries. As a point of contrast, Denmark provides displaced workers with 90% of their previous salary for four years.

In the United States, only one in three unemployed workers receives unemployment insurance. This is attributable, in part, to restrictions on who is eligible. Unemployment insurance generally does not cover temporary workers, the self-employed, agricultural workers, part-time workers, those who were out of the labor force for extended periods of time, those who voluntarily left their jobs, or those who are trying to reenter the labor force after an extended absence. It also fails to protect parents (most commonly women) who take time off to raise families. As a result, unemployment insurance covers only 40% to 60% of those who lost jobs, 10% to 15% of those who left their jobs, 25% to 35% of those trying to reenter the labor force, and only 10% of those who are attempting a new

Exhibit 5.2   Unemployment Entitlements: International Comparisons

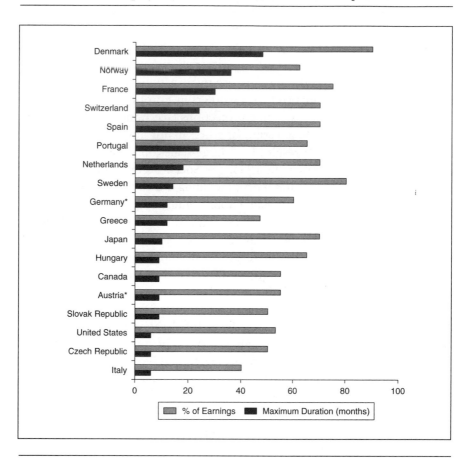

Source: Organisation for Economic Co-operation and Development (2004)

*Net earnings

entry into the labor force. Additionally, many workers do not apply for benefits because they lack an understanding of how to apply, assume that they will find replacement work soon, or feel a sense of stigma in using social assistance programs (O'Leary and Wandner 1997; Wandner and Stettner 2000).

In sum, job loss protections, developed during the 20th century in America, are meager, short-term, and restricted, leaving workers who lose their jobs with few resources upon which to fall back. This approach is increasingly out of step with the needs of workers laboring in an economy where job loss is commonplace.

How Insecure Are Workers in the New Economy?

In the old economy, the risk of unemployment was unevenly distributed. For those at the bottom, who worked in insecure secondary labor market jobs, bouts of unemployment were common and the risks of destitution were quite high. But for unionized workers, security was nested in seniority systems that ensured that the longer one worked for an employer, the lower the likelihood of job loss. Managers and professionals were protected by reward systems designed to retain highly skilled employees and reward organizational loyalty. Employers provided these protections not out of the goodness of their hearts, but because they were persuaded that it was in their own competitive interest to retain hard-to-replace employees and to invest in their professional development. Today, however, there is ample evidence that something has changed, so that even workers in "good jobs" are far less secure (Beck 2000; Rubin 1995; Rubin and Brody 2005; Smith 2002; Sweet, Moen, and Meiksins 2007).

How insecure are today's workers? The answer depends on how insecurity is defined. One way of measuring insecurity is to examine the percentage of workers who are **unemployed**—people who are actively looking for work but cannot find it. At any given point in 2005, nearly 10 million Americans—about 1 in 20 working-age adults—were unemployed. The rates for women and men were approximately equal, but African American men were unemployed at twice the rate of white men. Unemployment rates have fluctuated considerably during the past 40 years, from a low of 3.5% in 1969 to a high of 9.6% in 1983. These rates also vary monthly, and within industries. For example, construction workers are commonly laid off during the winter months, but retail employment opportunities grow in advance of the holiday season.

The relatively low unemployment rates in the contemporary United States might lead one to conclude that job insecurity is not a major (or growing) problem. However, relying on unemployment statistics as the exclusive measure of job insecurity tells only part of the story. According to the standards used by the Bureau of Labor Statistics, to be classified as "unemployed," an individual would have to be previously employed, not currently working, *and* actively seeking a job. As a result, these statistics exclude **discouraged workers** (people who have given up finding work), those who have been out of the labor market and who are actively seeking to reenter it, those who are forced to accept jobs that are inappropriate to their qualifications and needs, and those laboring in jobs that offer no protection beyond short-term contracts (Cottle 2001; Kalleberg, Reskin, and Hudson 2000; Smith 2002; Thurow 1992, 1999). One illustration of the limitations of these figures can be observed by examining job creation data, also collected and reported by the Bureau of Labor Statistics. Interestingly, when

new jobs are created, the numbers of unemployed do not go down propor-
tionately (Thurow 1999). The logical conclusion is that unemployment statistics fail to recognize many potential workers, people who could be
working but are not (Jahoda 1982; Kelvin and Jarrett 1985).

It should be emphasized that unemployment statistics also do a poor job
of revealing the number of people who lose their jobs involuntarily. As
Jacob Hacker (2006) has recently argued, the unemployment rate has not
grown much in recent years, but rates of involuntary job loss *have* grown
and have reached levels usually associated with a significant economic
recession. He concludes that many people who lose jobs find replacement
jobs that are inferior to their previous positions. These new jobs may offer
pay and benefits that are lower, and may be mismatched with workers' education and experience. As a result, these workers remain employed, but still
fear (and experience) employment insecurity (see also Kalleberg 2007).

Most workers are let go one by one, in numbers too small to register as
events worthy of a newspaper lead story. But in the new economy, these
trickles of job loss combine to form rivers of insecurity. Other workers are
let go together with sizable numbers of other employees in plant closures.
Beginning in the mid-1990s, the Bureau of Labor Statistics began tracking
mass layoffs, events in which establishments had 50 or more employees
applying for unemployment insurance within a 5-week period. Exhibit 5.3

Exhibit 5.3   The Impact of Mass Layoffs on Unemployment: United States
1996–2004

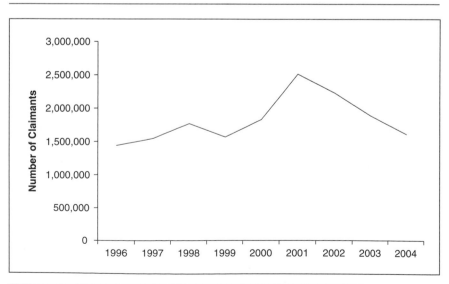

Source: Bureau of Labor Statistics

shows that in 2004, at least 15,980 establishments had large-scale layoffs affecting more than 1.5 million workers. The three most common reasons for mass layoffs were seasonal work ending, company restructuring, and permanent plant closure. Although reliable data on mass layoffs are not available for a long-term analysis, the number increased significantly in the early 1970s, particularly in the manufacturing sector, and has subsequently remained consistently high (Bensman and Lynch 1987; Bluestone and Harrison 1982).

The growth of job insecurity has been especially high among production workers as it spread beyond traditional secondary labor market jobs to highly paid factory workers in industries such as steel and automobile assembly. Here, the pressure to cut costs to deal with intensified competition, declining unionization, the ability to relocate production, and managerial strategies emphasizing downsizing and "lean and mean" organizational designs have had well-documented effects (Moore 1996). There is even evidence of mass layoffs in growth sectors of the economy. For example, Intel eliminated 10,000 jobs in 2006, and Hewlett-Packard cut 15,000 in 2005. Overall, information sector employment declined by 17.4% between 2001and 2006 (Wypijewski 2006). The question remains, however: Has job insecurity spread to other sectors of the workforce?

This has been the subject of much dispute among social scientists. Some argue that the effects of downsizing and mass layoffs have really been felt only among manufacturing workers; other groups remain relatively unaffected (Baumol, Blinder, and Wolff 2003). Skeptics note that data on job tenure (how long people have held their current job) do not show a clear, unambiguous drop in recent decades (Capelli 1999; Jacoby 1999). Even the much ballyhooed American Management Association report indicating that there had been significant layoffs among managers in the late 1980s and early 1990s has been questioned. The economist David Gordon (1996), for example, noted soon after this announcement that most displaced managers seemed to be finding new jobs and that the ranks of management in most companies were growing, not shrinking.

The skeptics are probably right to suggest that job insecurity among managers and professionals remains lower than among those in manufacturing and secondary labor market jobs. The reality is that hourly workers in most industries, skilled or unskilled, unionized or not, can no longer count on the security provided by large employers and seniority rules. Nevertheless, although some employers continue to place an emphasis on retaining high-skill employees (e.g., Microsoft), many employers have responded to increased economic competition and unstable market conditions by weakening their commitment to reward structures designed to encourage employee loyalty (Moore 1996). Thus, efforts to promote "flexibility" encourage looser commitments to all employees. The flexibility push is reinforced by stockholder pressures to engage in periodic bouts of downsizing and layoffs, by new forms of work organization that assign managerial tasks to worker groups, and by new organizational structures that make it harder to advance careers by moving up within a single employer. Additionally, fiscal pressure has increased insecurity among the many professionals employed in the public sector (such as teachers). Although some managerial and professional workers have done well in the new economy, it is also apparent that at least some of these once-secure workers cannot be as confident that they will retain their jobs and have to think seriously about the possibility of midcareer changes (Capelli 1999; Jacoby 1999; Moen and Roehling 2005).

Moreover, the experience of employment insecurity among these once-sheltered employees is not limited to the threat of job loss; it extends, too, to the economic security that traditionally went along with their kinds of work. Managerial and professional employees once could count on generous benefits packages and stable, steadily growing incomes. The evidence is that they have good reason to feel less confident that these arrangements will be available.

Hacker (2006) notes that American society has increasingly shifted "risk" onto the shoulders of employees. Instead of defined benefit health care plans, for example, employees are encouraged to open health care savings plans (which means they risk not saving enough). Instead of traditional pensions, employees are offered stock ownership plans, investment opportunities, and other market-based opportunities that promise greater payouts, but also involve the risk of loss. And those risks are real. This lesson was learned by Enron employees who lost nearly all their retirement savings after they followed advice to invest in company stock, even as their employers knew its value was plummeting. They were not alone; the 120,000 employees at United Airlines saw most of their retirement savings disappear when the company declared bankruptcy in 2004 (Marks 2004). Newspaper accounts of companies closing operations, underfunding their pensions (and petitioning the government for financial assistance), and engaging in legal maneuvering to get out of pension commitments to current and retired workers are now commonplace.

Income stability also has weakened in recent years. There have always been people who have suffered sudden, temporary drops in income (because of illness, job loss, business failure, etc.). But the numbers of people who suffer sudden losses in income have grown significantly. Moreover, the magnitude of their income fluctuations has also grown—Americans' chance of a 50% or greater drop in income rose from just over 7% in 1970 to almost 16% in 2002.[1] And, although the risks of such income fluctuations are still greater among the less educated, the risks grew more rapidly for better-educated Americans in the 1990s (Hacker 2006). With budgets stretched to the breaking point, for many dual-earner couples, an unexpected event such as a job loss or illness can plunge a family into bankruptcy (Warren and Tyagi 2003).

There is also a subjective dimension to job insecurity, the anxiety felt at the prospect that one's own job *might* disappear (Schmidt 2000). The General Social Survey, a widely used national opinion poll, indicates that feelings of job insecurity have been rising, especially among managers and professionals. In 1977, 1 in 20 (6%) managers and professionals reported that it was either "very likely" or "fairly likely" that they would lose their jobs within the year. By 2002, the proportion had doubled, with 1 in 10 (11%) reporting that that there was a serious possibility that they would lose their jobs. Another survey, the Couples and Careers Study conducted from 1998 to 2000 by the Cornell Careers Institute, surveyed dual-earner middle-class workers in upstate New York. This study revealed that most higher-level workers were only about 80% to 85% confident that their jobs would exist for the next 2 years. Only 1 in 4 (27%) men and 1 in 3 (35%) women in this study were fully certain that they would be able to keep their jobs (Sweet, Moen, and Meiksins 2007).

Finally, job insecurity is expanding because of the changing composition of the labor force, as well as because of the changing organization of work. In an economy dominated by dual-earner families, the loss of one spouse's job creates strong prospects that the other partner—even one with a secure job—may need to reconsider career options. Those partners can be put into the position of **trailing spouses,** and be forced to rework their careers to match the direction charted by their partner. Alternatively, displaced workers could have spouses who hold jobs that they are reluctant to relinquish. In this circumstance, displaced workers are anchored to communities, and their options can be limited. To fully understand the realities of job insecurity, therefore, requires thinking of *families* of jobholders as well as individual jobholders.

To highlight how this operates, let us reexamine the data from the Cornell study from the perspective of couples, and link spouses' careers together as they actually are in the lives of most Americans. Recall that the typical professional worker is about 80% confident that his or her job will remain for another couple of years. Using this as a dividing line, we can consider those at or above the 80% mark as holding comparatively "secure" jobs, and those below as being insecure. Remember that these workers are married to spouses who also hold jobs that may be secure or insecure. What are the prospects of both partners in a dual-earner family holding secure positions at the same time? Exhibit 5.4 reveals a sobering state of affairs in America—*secure working conditions are the exception, rather than the rule,* for dual-earner professional couples. Only 44% of American dual-earner households have both partners feeling fairly secure in their jobs; most families have one or both partners feeling insecure. If we raise the threshold to *full* confidence that jobs are completely safe, the numbers plummet even further, as only 1 in 10 (12%) of dual-earner households has both partners feeling fully secure (Sweet, Moen, and Meiksins 2007).

Exhibit 5.4   Job Security Configurations of Dual-Earner Professional Couples (80% Confidence That Jobs Will Be There in 2 Years)

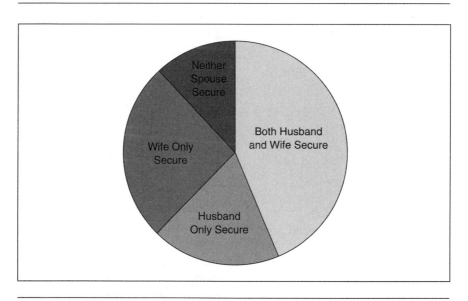

Source: Sweet, Moen, and Meiksins (2007). N = 125 Couples

The Costs of Job Loss and Insecurity

Losing a job has immediate and long-term consequences for household finances (Broman, Hamilton, and Hoffman 1990; Hoffman, Carpentier-Alting, Thomas, Hamilton, and Broman 1991; Leana and Feldman 1992; Sales 1995). Even among displaced professional workers, two in three receive no severance pay after being terminated, and for those who do get severance packages, the compensation varies greatly. Most receive only a modest sum of money (2 weeks' pay being the most common), providing little relief from the financial strains of finding new work (Sweet, Moen, and Meiksins 2007).

The economic impact of losing a job is not a new concern. What has changed in the new economy is the ability of families to weather these disruptions. In the old economy, the husband/breadwinner arrangement made it possible for wives to act as financial reserve units, and to hold jobs temporarily when their husbands lost work. Today, the incomes of both spouses are essential to meet family budgets. As a result, short periods of joblessness can devastate family finances in ways that were less common in the old economy (Moen and Roehling 2005; Warren and Tyagi 2003). In our own studies, when we asked displaced workers and their spouses what they wished they had done differently, the most common responses were to have saved more and to have put more energy into finding new work before losing their old jobs. But when families are putting in long, stressful hours of work already (see Chapter 6), this is often difficult to do. As one displaced engineer, a mother of an 8 year-old boy, told us,

> When I think of myself at that time [before I lost the job] . . . I felt like I was doing just about all I could do.[2]

Another problem facing insecure workers is that even though jobs are unstable, they remain tightly tied to individuals' senses of identity and feelings of self-worth. As a result, losing a job can have a major impact on health and social-psychological well-being (Burke and Greenglass 1999; Dooley, Catalano, and Wilson 1994; Hamilton, Broman, Hoffman, and Renner 1990; Iversen and Sabroe 1988; Jahoda 1982; Kelvin and Jarrett 1985; Leana and Feldman 1992; Perrucci, Perrucci, and Targ 1997; Shamir 1986a; Uchitelle 2006; Vosler and Page-Adams 1996). Job losses also strain marriages, sometimes to the breaking point (Perrucci 1994; Shamir 1986b; Westman, Etzion, and Danon 2001). Consider, for example, how job loss affected Lisette, a displaced 48-year-old clinical nurse, and her relationship with her husband:

> [My boss] just told me that she had some bad news. That, you know, that they had to make the budget cuts. And my position was one of the positions that was being eliminated. And, I mean, I just burst into tears . . . And I was just so shocked. Everybody that worked with me was shocked. Just, um, it was just really unexpected and nobody had thought that, you know, it would happen to somebody like me, you know, who had done such a good job and had worked so hard and I was really well liked by everybody there.

Feeling "lost," and perhaps in a state of denial, Lisette had difficulty understanding the permanence of her employer's decision and returned to meet with her boss in the following week to negotiate a means to get her job back. In shame, she avoided telling her husband, Damon, about the job loss, and for an entire week she pretended to go to work. Damon, in turn, interpreted this as a "personal rejection," and their mutual satisfaction with their marriage dropped precipitously.

Interestingly, it is not only those who lose jobs and their spouses who are negatively affected. Simply witnessing coworkers being displaced increases anxieties and stress in the workplace (Armstrong-Stassen 1998; Brockner 1990; Brockner, Wiesenfeld, and Martin 1995; Brockner, Grover, Reed, DeWitt, and O'Malley 1987; Grunberg, Anderson-Connolly, and Greenberg 2000; Kets de Vries and Balazs 1997). However, numerous studies show that this stress can be moderated if employers make visible the rationale for organizational restructuring, provide advance notification that job loss may occur, and give evidence that procedural justice has been used to determine who will (and who will not) lose jobs (Appelbaum, Simpson, and Shapiro 1987; Armstrong-Stassen 1993; Brockner 1990; Latack, Kinicki, and Prussia 1995; Mishra and Spreitzer 1998; Naumann, Bennett, Bies, and Martin 1998; Wanberg, Bunce, and Gavin 1999; Wiesenfeld, Brockner, and Martin 1999). In other words, if workers know that jobs have to be lost, agree that job losses are necessary, and are given opportunities to plan, they are in a stronger position to adjust their lives accordingly.

Psychologists have long documented how planning and a sense of control facilitates social psychological adjustment to change (Bandura 1982a, 1982b). Toward this end, advance notification can be the key to helping workers adjust *in anticipation* of job loss. If workers know ahead of time that layoffs are being planned, they are in a position to seek additional training well before they find themselves without a job. In addition, advance notice can give workers and their representatives time to negotiate with employers either to limit the numbers of people affected or to arrange for assistance of various kinds for those employees who are displaced (Moore 1996; Uchitelle 2006).

Outside of individually negotiated contracts and collective bargaining agreements secured by unions, the only "right to know" about job futures is in the **Worker Adjustment and Retraining Notification (WARN) Act.** This law mandates that companies with more than 100 workers provide 60 days' notice if they anticipate terminating or laying off 50 or more employees. Though of significant benefit to workers in large companies, this act leaves most workers unprotected—those employed in smaller companies, those who are not part of a mass layoff, those who have worked for the company less than 6 months, and part-time workers.

Unfortunately, most employers are reluctant to extend advance notification of impending job loss. Exhibit 5.5 shows that more than half of the displaced workers we interviewed said that they received no notification, formal or informal, that their job was going to be eliminated. Only a small minority, one in five workers, had 3 or more months to prepare for the loss of their jobs. And even when employers extended severance packages, the callous methods used to inform employees of layoffs remained a bitter pill. Our in-depth interviews with managers and skilled professionals revealed sobering accounts of what it is like to lose a job in the new economy. For example, Edwin, a 40-year-old engineer and father of three young children, told us of how he lost his job:

> How did I learn my job was gone? Um, just all at once. I had no idea I was going to be let go. I was actually saying goodbye to people that were being let go. And, you know, kind of comforting them and telling them it would be okay. And I was sitting at my desk later that afternoon and I just see my boss walk up to me and ask me if I had a minute. And I kind of saw it in his face. And I say right then, "You got to be kidding." He says, "Yup, sorry, bad news." So that was it. It was that cold.

Exhibit 5.5   Percentages of Displaced Professional Workers Who Received Notification That Their Jobs Would Be Eliminated

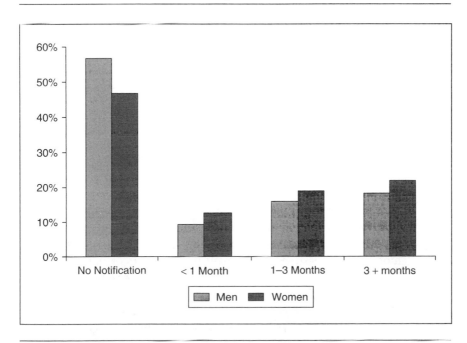

Source: Authors' Analysis of the Couples Managing Change Study. N = 83.

This case is typical. Before being let go, Edwin and his wife had it all—two good jobs, a satisfying family life, and a happy marriage. Their success came after years of working together to locate two jobs, select their community, and reconfigure their work schedules to raise their children (Moen and Sweet 2003; Pixley and Moen 2003; Sweet, Swisher, and Moen 2006). In the course of an afternoon, their lives were turned upside down, forcing them to confront difficult, fundamental decisions: how to manage dwindled resources, whose job to favor in the next stage of Edwin's career, and how to cope with the prospect of losing their home and neighborhood ties.

Even when employees have some advance knowledge of impending layoffs, the consequences of layoffs can be significant. This is particularly true when layoffs affect declining regions, single-employer communities, and older workers (Sweet 2007). In these situations, workers often have great difficulty finding new jobs because there is little else available locally to employ them. Younger workers may be willing to move away, but older workers typically are more reluctant and are more likely to experience prolonged or even permanent unemployment. The result can be both economically and psychologically devastating, undermining a worker's sense of self as a breadwinner and causing workers, both young and old, to look at work more as a way to pay the bills and less as a source of personal satisfaction and community (Koeber 2002).

Responding to Insecurity: Old and New Careers

The old economy encouraged an orientation to work and careers that is discredited in the new economy. For those in the primary labor market, a "good employee" was someone who had found a well-paid, stable job and who displayed loyalty to the organization. Jumping from job to job tended to be seen either as opportunism or as a sign of a troubled work history. For those in the secondary labor market, unstable work histories reflected career patterns that were fostered by limited opportunity, but were also used as evidence of individuals' unsuitability for jobs that required greater responsibilities. As a result, demonstrating a lasting commitment to employers or unions became essential for upward mobility, and an employee's loyalty became synonymous with her or his "character." To prosper as an "organization man" or a "union man" required embracing an employer's or union's interests as one's own, and seeing oneself as part of an association of like-minded employees (Mills 2002 [1951]; Whyte 1956).

For those in primary labor market jobs, employee loyalties in the old economy were commonly reciprocated by employers, who adopted an ethos of **corporate paternalism**—the belief that it is the employers' responsibility to provide for their employees, much as it is parents' responsibility to provide for their children. For instance, in the 1950s, IBM CEO Thomas Watson Jr. proudly stated, "There are many things I want this company to become, but no matter how big we become, I want this company to be known as the company that has the greatest respect for the individual." Consistent with Watson's words, IBM extended generous pay, lifetime employment, and jobs to widows (Kanter 1977).

In the latter part of the 20th century, the CEO's role as the benevolent employer was replaced by a new kind of corporate leader, exemplified by Sunbeam CEO Albert "Chainsaw Al" Dunlap[3] and the legions of corporate raiders who made quick profits by dismantling otherwise profitable enterprises (Barlett and Steele 1992). Perhaps the most vivid portrayal of the new corporate ethos is offered in the Michael Moore film *Roger and Me,* which shows General Motors CEO Roger Smith dodging the question of why American jobs were being moved offshore. In contrast to Watson, Smith defined his responsibility as being solely to the investors of General Motors.

One of the reasons why jobs are insecure in the new economy is that there is a growing acceptance—especially among managers and captains of industry—that jobs *should be* insecure (Thurow 1992). As Brian O'Reilly stated in a widely cited article in *Fortune* magazine, the "new deal" for workers is this:

> There will never be job security. You will be employed by us as long as you add value to the organization, and you are continuously responsible for finding ways to add value. In return, you have the right to demand interesting and important work, the freedom and resources to perform it well, pay that reflects your contribution, and the experience and training needed to be employable here or elsewhere. (O'Reilly 1994)

O'Reilly was writing to the select workers (and their employers) who benefited under the old system of rewarding seniority with job protections. He presented a case for the continual restructuring of work, fitting people to jobs (as opposed to the reverse), and the dismantling of jobs when work is completed. His view reflects the themes of "rightsizing" and "organizational flexibility" that were emphasized in numerous managerial publications of the past 30 years. The accepted goals of modern business are to be "lean and mean" and to restructure continually—even as a number of critics argued that this often has negative effects on corporate profitability in both the short and the long term (Grunberg, Anderson-Connolly, and Greenberg 2000; Sun and Tang 1998; Wagar 2001). Although the new social contract has been packaged as providing flexibility, from a workers' perspective it can often look much more like an effort to extend the power to displace employees at will (Pollert 1988).

The problem of job insecurity is pervasive in secondary labor markets, where workers enter jobs with virtually no protection against dismissal. For example, an application for work at a Burger King restaurant informs workers that they "can resign or be terminated . . . at any time without notice or requirement of cause." Although their job application form states that they have access to an arbitration program, there is virtually nothing to arbitrate if the signed agreement says that employers can fire workers at will (Wolkinson and Ormiston 2006).

Fast-food workers are not the only ones who experience this one-sided arrangement. Our interviews with displaced professional workers revealed many stories that show how little loyalty is extended from the top down. Even dedicated, long-service employees are displaced at will, and often in ways that show little or no appreciation for employees' past contributions. The account offered by Duke (a 36-year-old engineer at a major manufacturing company) was similar to those of many other workers who had lost their jobs:

> They had everybody in the organization go to a meeting. They handed out paper slips, some of the people went upstairs, some of the people went downstairs, and the people who went downstairs were informed that they were done as of that day, and the people who went upstairs were informed that they were going to be kept on for a couple months but their jobs would also be gone . . . It was a real lousy way to do something like that.

As we made comparisons between workers and looked at how employers let workers go, we found that Duke and his coworkers actually reported comparatively favorable treatment, in that they received 2 weeks' severance pay for each year they had been with the company, as well as temporary continuation of their health insurance and access to an outplacement company. In contrast, Tobias, a 53-year-old information specialist, recounts a worse and sadly common experience. After working a series of 12-hour days to complete an important contract, he was called into his supervisor's office, told that his job was terminated, and was escorted out of the building a few minutes later. Because of fear that he would sabotage work stored on his computer, he was not allowed to log back into the system or remove any personal communications or files on his machine. He was not alone. In the weeks that followed, 12 other employees at his company were treated in a similar manner. Although

Tobias' story is particularly harsh, he is among many others we interviewed who were "tapped on the shoulder" and escorted out of their buildings within a half hour of being notified that their jobs were gone.

To what extent is the new social contract reshaping workers' commitments to their jobs and employers? Some have argued that the new economy is corroding the character of American workers and undermining lasting commitments to fellow workers and employers. This portrait of the modern worker depicts her or him as pursuing immediate self-interest, but lacking a commitment to labor for a greater good. Like their employers, these workers have abandoned the values embraced in the old economy and no longer see loyalty as a virtue (Sennett 1998).

Although this may be true for some employees, our interviews with displaced workers did not indicate this to be as common as is often argued. Instead of being disloyal, we found that many of the problems experienced by displaced workers resulted from their being *too loyal* to their old employers. They were intensely committed to their jobs and worked long hours not only because they had to, but because they saw it as their duty. For the most part, they continued to behave according to preexisting models of what a good employee should be and to give as much as they could to their employers. As a result, the most common response we heard from workers (even those who had a sense that their jobs were unstable) was that learning of their job loss was like "being a deer in the headlights," "being hit on the head," "stunned," "shocked," "distressed," "incredulous in disbelief." Had these workers constructed their value systems and personal strategies to correspond with the new economy, they would not have reported that "employers do not take care of their people," and that they felt "unappreciated," "annoyed," "miffed," "frustrated," and "extremely angry." Workers adopting a new set of work values would be much more likely to anticipate job loss and be better prepared to cope with it both practically and psychologically.

Still, workers' orientations to jobs and careers may be slowly changing. The Bureau of Labor Statistics reported that in 2004 the average length of time workers had been with their current employer was just 4 years. The average person in the United States will hold nine jobs between the ages of 18 and 34, and several more during their careers. More importantly, far fewer workers expect, or plan, to stay with their current employers throughout their career, a sharp departure from the old ways of working. These data suggest that workers may be redefining career paths, as well as their strategies for navigating labor markets over the life course.

Some have argued that redesigned workplaces offer some segments of the workforce expanded possibilities to develop skills. Although these skills may not ensure security in internal labor markets, they are transportable to other places of employment. This optimistic view envisions a mobile workforce charting **boundaryless careers,** engaging in transient relationships between different employers and thereby expanding a variety of portable skills (Arthur and Rousseau 1996; Raider and Burt 1996). Most analysts who adopt the phrase "boundaryless career" use it to emphasize the ways in which individual choices are expanded in the new economy. Such is the case for computer programmers in Silicon Valley. As these workers rotate through a variety of companies, they gain a variety of experiences and earn significant financial rewards, simultaneously enhancing innovation and expanding knowledge (Saxenian 1996). However, the cavalier use of the word "boundaryless" to describe careers needs to be questioned when considering the constraints on worker mobility. For some workers, especially single workers, movement from job to job or community to community is relatively unconstrained. But boundarylessness ignores the everyday constraints on workers that stem from their relations with others. Relocating, for most workers, entails considering how a career move may affect the career of a spouse, their children's well-being, and the ability to care for aging parents. Rather than unconstrained careers, workers in the new economy face a variety of new constraints—in terms of social relations, economic resources, and opportunities.

The problem of career instability is also apparent in worker experiences in retraining programs. Many laid-off workers are encouraged to seek additional training, either by going back to school, or more often, by taking advantage of either publicly funded or employer-sponsored retraining programs. Rather than remaining committed to their old jobs (and hoping for a recall or a new job just like their old one), workers are told that they should retool, acquire new skills, and adapt to changing patterns of demand for workers. Unfortunately, worker retraining often produces cynicism and disappointment, rather than employment. It is hard to know *for what* to retrain workers (because predicting employment demand is notoriously difficult); retraining often emphasizes short-term skill acquisition, rather than the broad education that would make employees truly "flexible"; and retraining often winds up training people for jobs that don't exist. Displaced workers are often reluctant (quite understandably under the circumstances) to abandon their long-standing ties to occupations and employers (Moore 1996; Uchitelle 2006).

Additionally, the focus on the upward mobility of unattached professionals pursuing "boundaryless careers" ignores the careers of workers who labor in less favorable arrangements. A large proportion of the workforce labors in **contingent jobs,** "as-needed" positions, filled by short-term agreements between workers and employers. These jobs include on-call work, temporary help agency work, and contract work positions designed to be temporary and unstable (Polivka and Nardone 1989). According to the Bureau of Labor Statistics, in 2005, there were three times more temporary agency workers than there had been in 1990, and these jobs are most commonly occupied by women, younger workers, and members of minority groups. This growth corresponds with a growing reliance on **contract workers,** individuals employed for fixed terms with no promise of future employment.

Because of their short-term employment arrangements, contingent workers are commonly treated differently than "regular workers." For example, temporary workers are sometimes given different-colored security badges, assumed by coworkers to be less competent, not offered employer-sponsored training, and excluded from office rituals such as birthday parties. Because of their perceived marginal status, coworkers are less inclined to express interest in these workers' lives outside of the workplace, and often fail even to learn their names (simply referring to them as "the temp") (Henson 1996; Smith 2002). A widely cited study found that contingent workers earn considerably less pay, have less access to health insurance, and are less likely to receive pension benefits than those who work in more conventional arrangements (Kalleberg, Reskin, and Hudson 2000). Temporary workers sometimes report liking aspects of their jobs, but as Vicki Smith (2002) forcefully demonstrates in her book *Crossing the Great Divide,* nearly all long for secure jobs, something harder to find in the new economy.

Conclusion

In this chapter, we considered risk and job insecurity for American workers in historical and international comparative contexts. We showed that the element of risk is not new and, in many respects, workers today are far better off than were workers in early phases of industrialization. But new problems are emerging, including an expanded and broadening spectrum of jobs that offer little security. Contemporary employees face different types of risk than were experienced by employees in the old economy, not only in their employment contracts, but also in the ability of their families to weather career disruptions. And the demography of the American workforce has changed, not just jobs. Dual-earner families are in a unique position of double jeopardy because the loss of one partner's job can disrupt both partners' careers.

Unlike their European counterparts—who are protected from risk simply by virtue of citizenship—American workers have few resources to protect them from the consequences of job insecurity and loss. American workers need to hold jobs, but also to find jobs that offer protections in the form of "benefits." When these jobs are lost, they need to find replacement work quickly. The safety nets established during the 20th century were built to shelter husband/breadwinner–wife/homemaker families through brief periods of joblessness. These risk management strategies are now woefully insufficient to protect most working families, who need dependable incomes from two workers to make ends meet. Similarly, providing pensions and other benefits through employers has become more and more problematic, as workers are forced to move laterally from job to job and from employer to employer.

The new economy has fundamentally weakened the social contract between employers and employees. Employers once believed it was in their own interest to retain valuable employees, but economic instability, global competition, and stockholder pressure have encouraged them to feel free to move workers out of jobs at will. This new managerial ethos enables employers to serve the interests of stockholders through flexible labor management strategies, but at great cost to American workers, who often are unsure if they can depend on employment beyond the short term. And because employers seldom provide employees with solid information about planned layoffs, workers typically lack the ability to gauge their future job prospects accurately. A more equitable arrangement would necessarily involve providing workers the opportunity to plan careers proactively in an economy of expanded risk, and to make tactical career decisions based on solid information about their future employment prospects. Instead, most workers are obliged to be reactive, to redirect disrupted lives in response to decisions they had no part in making.

Notes

1. These data are drawn from Hacker's (2006) analysis of data from the Panel Study of Income Dynamics, pp. 30–31.

2. Quotes in this chapter are taken from the Couples Managing Change Study (Stephen Sweet and Phyllis Moen, coprincipal investigators), in-depth interviews with 125 dual-earner couples who either held insecure jobs or who had recently lost jobs. For details on methodology see Sweet, Moen, and Meiksins (2007).

3. This nickname was originally coined by workers protesting his labor management practices, but was also one that he embraced.

6

A Fair Day's Work?

The Intensity and Scheduling of Jobs in the New Economy

The workers we described at the beginning of this book face a number of challenges in managing the time and intensity of their work. Eileen (the engineer) was able to negotiate a schedule that enabled her to reduce the time she spent at work, but this contributed to her falling off the career track. To get her career back on line required her again to commit to putting in long hours of work, at considerable expense to her son's needs. Dan (the displaced autoworker) went from a job with a predictable schedule to self-employment, where he could determine his own hours. For this worker, self-employment created the freedom to fashion a schedule around his other interests, but left him feeling insecure about where the next dollar will come from. Jamal (the fast-food worker) found himself working a job with an undependable schedule that left him short on hours and income. And, as anyone who has ever worked in a fast-food restaurant knows, this kind of work provides few opportunities to rest because there is always something that needs cleaning, moving, or preparing. The job demands that Chi-Ying is expected to satisfy in her factory in China are even more intense. Like her coworkers, she commonly works 6 to 7 days per week, longer days than American workers typically work, and she is expected to do so at a pace that might best be described as frantic.

All of these workers face common problems—negotiating work arrangements that enable them to meet their family needs, keep their careers intact, and gain a fair day's wage for a fair day's labor. What is a reasonable workday for an employee in the new economy? And should employees be expected to perform—as Frederick Winslow Taylor (1964 [1911]) prescribed in his *Principles of Scientific Management*—the absolute maximum amount of labor possible, or should their work occur at a more relaxed and (some would argue) humane pace?

This chapter focuses on the intensity and scheduling of work to understand the challenges that job expectations pose for workers and their families in the new economy. We consider the factors that cause some workers to labor longer than they want to, while (ironically) others work less than they want to, or even not at all. We ask why so many Americans feel overworked. We place job demands in historical and comparative perspective to highlight the ways in which American culture and policy contribute to work-poverty and overwork. Finally, we consider the impact new schedules have on working families and their strategies for reconciling job schedules with family demands.

Time, Intensity, and Work

One obvious way to measure how much work people are performing is simply to count how many hours they spend laboring. Interestingly, this way of thinking about work commitments is relatively recent. Only in the last two centuries have jobs become things people "clock into." By contrast, in preindustrial times, work centered on the home and constituted a stream of activities that varied according to the rhythms of the day and the season. Workers responded to (among other things) cows returning from the field, hay in need of sowing or harvest, and children in need of nursing (Osterud 1987). The labor historian E. P. Thompson (1967) termed this **task-oriented labor**—a systemic approach that organized, measured, and compensated work on the basis of the completion of specific assignments. When work was organized according to task, schedules varied remarkably throughout the year and typically interspersed intense labor with prolonged periods of relaxation. And because work was performed in and around the home, people in preindustrial societies did not make a clear distinction between what was "work" and what was "life" because the two were inextricably intertwined.

Industrialization sparked a radical reorganization of work, shifting the criteria for compensation away from the completion of tasks in favor of the *length of time* engaged in tasks. This system of **time-oriented labor** led to the development of hourly wages and shift work and reoriented the rhythms of work and family lives around clocks and calendars (Hareven

1982; Marglin 1982; Thompson 1967). Although the hourly wage is now a taken-for-granted arrangement, in the early 19th century, this was a radical new practice that was not immediately embraced. Workers had to learn to arrange their schedules to match that of the factory and structure their family lives to perform shift work. As Thompson (1963) noted, the end of the Industrial Revolution can be marked by the moment workers stopped protesting over *how much* they were expected to produce and started protesting over *how long* they were expected to work—a change that transpired around 1830. The factory bells that echoed through New England in the 19th century continue to resonate today. Many taken-for-granted divisions of the day, week, and year—including vacations, the weekend, and the lunch hour— were all literally invented in the late 19th century and reflect the triumph of time-oriented systems of controlling work (Rybczynski 1991).

Arguably, a task orientation to work is a more natural means of organizing work than is a time orientation. Consider, for instance, that when college students are assigned to write a term paper, their most common question is "How long does it have to be?" Students never ask, "How much time

Exhibit 6.1   The Time Clocks That Track Worker Effort Remain a Legacy of the Old Economy

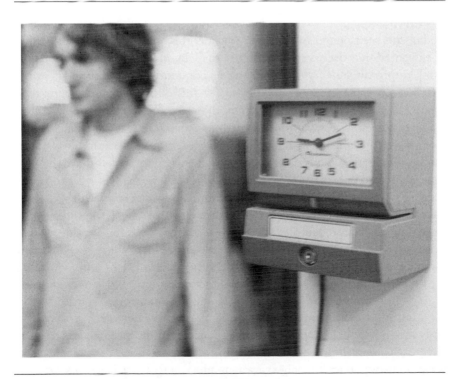

will I be required to spend writing?" The first question is more natural because it concerns the product of one's efforts. The latter question is of greater concern for most workers in the labor force because their compensation is not determined by how productive they are, but by how long they spend engaged in the activity. Today, even some skilled knowledge work is measured by the hour. For instance, lawyers seek to maximize their "billable" hours because this constitutes the measure by which they are commonly paid (Yakura 2001). Although progress would suggest that work would have grown easier, more fulfilling, and more economically rewarding, the 60-hour workweeks of the 19th century have not faded entirely away. Many Americans find themselves working longer hours and harder than they want to, and more than their parents did. The hope that a new economy would enable people to work less, not more, has thus far been disappointed (Schor 1991).

Understanding time-oriented labor as a socially negotiated system of managing work presents interesting and important questions about what activities should be compensated in the new economy. Many of the problems workers confront today emerged from the ways the terms of labor came to be negotiated in the old economy. For example, one old question concerns when work begins and ends. When miners and mine owners in the late 19th century faced this question, miners ultimately were forced to agree to walk or crawl several miles to reach the coal face before they were able to clock in to their jobs; paid work commenced only after they had begun to dig.[1] Today, this seems unreasonable, but consider this fact: The 2000 U.S. Census documented that the average worker in the United States spends 50 minutes a day (approximately 200 hours in a year) commuting to and from his or her job, but few workers are compensated for this effort. Acknowledging this, however, would require developing mechanisms for rewarding labor in all its forms, including work in and around the home, and replacing the restricted definitions of work created in response to industrialization.

The fact is that many jobs today require workers to labor before they enter and after they leave the workplace. Workers in many service jobs, for instance, commonly have to perform additional work tasks, such as cleaning their uniforms, on their own time. The time and expense it takes to be educated and trained for skilled jobs in the new economy is rarely included in calculations of worker effort. As a result, many young workers enter their careers burdened with tens of thousands of dollars of debt accrued during their college educations. This is a burden that they carry, not their employers.

As we discuss later, it is not simply working long hours that is the problem. Work in the new economy is also colonizing domains from which it was previously absent. Consider, for example, the extinction of the lunch hour, as well as the growing tendency to take work home on weekends, in the evenings, and along on vacations (or to forego vacations altogether). These behaviors were the exception in the 1950s; now they have arguably become the norm. Workers find that entering careers requires earlier demonstration of significant accomplishments and larger preinvestments in training than was the case for the previous generation of workers. College faculty offer a useful case in point, as junior faculty are now expected to publish multiple articles while pursuing graduate degrees and then to jump higher bars for favorable tenure reviews than their colleagues who entered the profession in the 1970s. Nor is it only workers at the high end of the human capital spectrum who need to make preinvestments to get jobs. A high school diploma used to be the requisite for a secretarial job, but those applying for administrative assistant positions today commonly compete with others who possess associate's and bachelor's degrees. And the drive to compete is extending downward through the life course, as evidenced by the intense pressure high school students experience to "stand out" among their peers. Even then, many of these students are finding that even a near-perfect performance provides no guarantee of access to the elite college programs that would have been a lock for high achievers of the previous generation (Rimer 2007).

Measuring the extent to which work has intensified is a challenging endeavor. One approach is to consider the average hourly output of workers, which the Bureau of Labor Statistics reports has consistently advanced during the latter part of the 20th century. Interpreting these data is complicated, however, because some of these increases can be attributed to technological or organizational advances. But even if this is the case (as we discuss later), these innovations have not reduced the burden of work. If anything, they have increased the polarity between those who have steady work (and are intensifying their efforts) and those who do not (for whom not working is hardly "leisure"). Indeed, there is a deep irony in the fact that the technological and organizational transformations of work that increased productivity have not resulted in an expansion of leisure.

How Long Are We Working? Comparative Frameworks

In the 1980s, pop psychologists recast the problem of overwork as a personal shortcoming akin to alcoholism. They argued that American society was inhabited by a growing number of "workaholics," people so addicted to their work that their jobs were their primary focus in life, which in turn caused their personal lives to suffer (Machlowitz 1980). As is the case with alcoholism, marking the line of excessiveness has proven to be a complex endeavor. Are Americans working too much? How do we know? From a sociological perspective, an even more important question emerges. If we are working too much, is it the result of personal shortcomings (as the pop psychologists suggest) or institutionalized arrangements? The answers to these questions require examining how much people are working, their own accounts of why they are working, and placing these findings in relation to work as it occurs in different types of occupations and in other societies.

One means of gauging American overwork is to examine how much work is being performed in other countries. Comparisons reveal that people in only a few countries labor as many hours as workers do in the United States. Exhibit 6.2 shows that Americans work, on average, 35 hours more

Exhibit 6.2   Average Annual Work Hours: International Comparisons

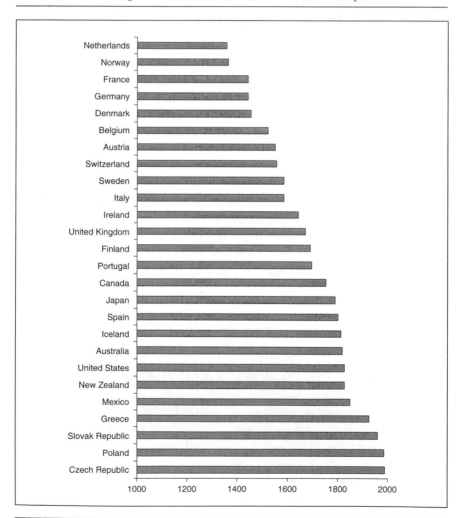

Source: Organisation for Economic Co-Operation and Development (2006)

Note: Hours represent data for most recent years available.

per year than Japanese workers (roughly 1 extra week), 155 hours more than British workers (roughly 4 extra weeks), 383 hours more than French workers (roughly 10 extra weeks), and 467 hours more than workers in the Netherlands (roughly 12 extra weeks). Although Mexico, Greece, and Poland all report higher average work hours than the United States, women in these countries are less likely to be in the labor force. Thus, the United States is exceptional in the amount of time its population devotes to work.

In comparison with most other countries, the United States is unusual in its lack of policies to limit the extent to which employers can overwork their employees. Two in three countries in the world have established laws to regulate the maximum number of hours employees can be expected to work (most commonly between 48 and 60 hours/week). Employees in most countries have a right to rest breaks, which in some countries are taken collectively, thus enabling nearly the entire society to pause from work. In the United States, no such laws exist at the federal level. In fact, laws and corporate cultures in the United States reinforce the *employer's* right to demand that workers engage in overtime work, even when that work directly conflicts with family needs (Jacobs and Gerson 2004; Wolkinson and Ormiston 2006).

The United States has fewer public holidays than most other developed countries and has no law guaranteeing its workers a right to vacations (McCann 2005). Exhibit 6.3 reveals that most European nations mandate that employers give workers 3 to 6 weeks of paid vacation a year. In contrast, the average full-time worker in the United States, though not entitled by law, gets about 2 weeks of vacation per year. Unionized employees receive more vacation, but with declining numbers of employees covered by collective bargaining agreements, the ability to secure vacation time has eroded (Mishel, Bernstein, and Allegretto 2005). Additionally, American workers fail to use all of their vacation time (or take work with them on vacation) for fear that they will fall behind in their jobs or careers.

Interestingly, Americans and Europeans took similar amounts of vacation in the 1930s, 1940s, and early 1950s. The subsequent divergence occurred because of different approaches to implementing vacation policy. In the United States, vacation leave was introduced in the 1930s and 1940s by employers, who regarded it as a means of increasing productivity by increasing employee satisfaction and health. It was also included in collective bargaining agreements, which expanded the amount of vacation time available to union workers. By contrast, most European countries introduced vacation time as part of government policy along with a host of other entitlements such as universal health care and job protections. Because of this, vacations are available to all workers in these countries, not just workers who have jobs that offer benefits or those protected by union contracts.

Exhibit 6.3   Statutory Annual Paid Vacation Days: International Comparisons

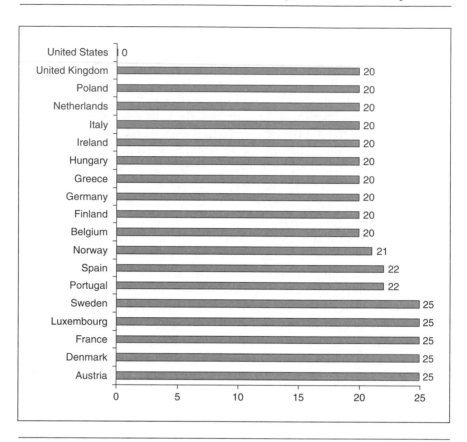

Source: Mishel, Bernstein, and Allegretto (2005)

If similar mandatory vacation laws were enacted in the United States, American workers would be laboring closer to the number of hours worked by Europeans (Altonji and Oldham 2003).[2]

Averages are informative, but equally important are statistics that document the diversity of work schedules within American society. According to the Bureau of Labor Statistics' (2007b) Time Use Survey, during their work-days in 2006, men spent on average 8 hours and women spent 7 hours laboring in their jobs. But as Exhibit 6.4 shows, fewer than 1 in 2 Americans worked the conventional full-time commitment of 40 hours a week. One in 4 Americans worked part time, and 1 in 4 worked more than full time. One startling fact is that nearly 1 in 10 workers commits 60 or more hours a week to their jobs. This is the group considered by pop

Exhibit 6.4   Average Hours Worked: United States 2003

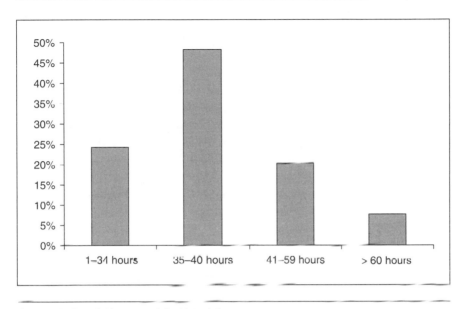

Source: Statistical Abstracts of the United States

psychologists to be workaholic. However, ethnographic accounts of lower working-class lives reveal that long hours often are not a choice, or the result of an "addiction" to the joys of cleaning floors and waiting tables. People in jobs like these work long hours because it is the only way to make ends meet (Ehrenreich 2001). Those in "good jobs," those working as managers and professionals, report that they would prefer to work fewer hours. Surveys show that a sizable proportion of the workforce (as many as one in five) would be willing to take a reduction in pay if it would enable them to work shorter hours (Galinsky and Bond 1998; Golden 2005). When highly committed workers are asked why they work longer than their ideal, the most common reason they offer is that "my job requires it" (Moen and Sweet 2003).

Analysis of data from the Current Population Survey (CPS) reveals that men's average work hours have not changed much during the past 30 years. What has changed is the likelihood that there will be two paid workers in the family and that women will be holding full-time jobs. These facts call for a shift in analysis, away from the time commitments of individuals to that of *households*. Jacobs and Gerson (2004) reveal in *The Time Divide* that the single biggest explanation for the expansion of work hours is the increasing prevalence of dual-earner couples.

Exhibit 6.5 illustrates the impact this has had on working families, showing the changes that occurred from 1970 to 2000 in the hours worked by husbands, wives, and couples. In 1970, the average wife worked only 12 hours per week, but by 2000, these hours had more than doubled.[3] The increase in women's paid employment was not associated with a decline in men's work hours, which also increased slightly during this period. As a result, in comparison with 1970, the "average" couple now puts 11 more hours into paid work each week. During this same period, there was little change in the total time spent on "home work," which in the 1980s (27 hours) remained approximately the same as in 1965 (24 hours) (Robinson and Godbey 1997; Schor 1991). As a result, considerable strains are created by dual-earner work arrangements, which often pit work commitments against family obligations, leaving workers—especially female workers—stressed and exhausted (Hochschild and Machung 1989; Moen and Roehling 2005; Pitt-Catsouphes, Kossek, and Sweet 2006).

Exhibit 6.5   Trends in Husbands', Wives', and Couples' Combined Weekly
Hours of Paid Work: United States 1970 and 2000

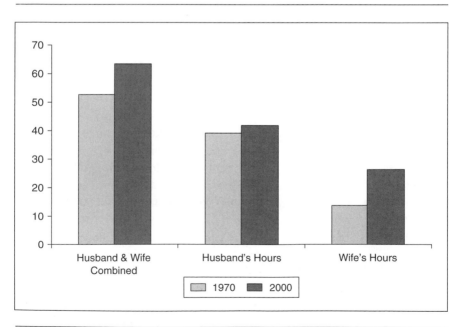

Source: Statistical Abstracts of the United States

In sum, American workers are laboring long hours. Many are working more than they did a few decades ago. Americans work more than workers in other developed nations, and often more than they want to. With this in mind, it is important to emphasize that many workers lack jobs, and many would like to work more than they currently do but have trouble finding enough work. During the 20th century, a divided economy emerged, one in which there is a **work-rich** labor force that labors long hours and a **work-poor** labor force that lacks sufficient work opportunities and compensation. Arrangements that compel some workers to perform the jobs of two workers (a product of both the number of hours and the intensity of work) block the creation of opportunities that could lead to employment for all. The problems of overwork and underwork, therefore, need to be viewed as two sides of the same coin (Gorz 1982; Jacobs and Gerson 2004; Rifkin 2004).

Finally, it is also important to remember that the reconfiguration of work demands is affecting developing societies as well. In factories in China, and in the *maquiladoras* that line the Mexican border with the United States, workers commonly labor 12-hour days, throughout the entire week. As factory work has expanded in these industrializing regions, rural migrants find themselves subject to the rigid, clock-driven rhythms of time-oriented labor. Although these labor practices make it possible for Americans to purchase inexpensive commodities for use in their scarce leisure time, they are a burden placed on other workers such as Chi-Ying. It is hard not to conclude that the new economy, in both its domestic operations and global operations, is compelling workers to labor more, not less, and expanding the chasms that separate workers from the opportunity to have a life off the job.

Working Long, Working Hard

Americans work long hours and work very hard when they are at work. Even some workers whose hours are not unusually long, and whose hours of work have not increased over time, complain of feeling overworked (Robinson and Godbey 1997). A recent Princeton Survey Research Associates study reports that 75% of Americans believe that workers experience more stress on the job than they did a generation ago (National Institute of Occupational Safety and Health 2007). Findings such as these indicate that the pace and intensity of labor and the duration of work are increasing.

Jamal's low-skilled restaurant job, like other forms of work designed in the old economy, were structured to pressure workers to "hustle" and "bull and jam." As many of these jobs remain in the new economy, new pressures have been introduced by new workplace practices and opportunity structures. For example, although smart technologies have eliminated some of the unpleasant, physically demanding tasks of the past, they also pressure workers to keep up with machines that work steadily and fast. Even workers whose jobs involve monitoring technology can experience high levels of stress because this type of work often involves a combination of repetitive tasks with the need to stay alert. To get a sense of these pressures, consider what it is like to be a baggage screener or ticket agent at a contemporary U.S. airport.

Although the collective pressure in the old economy was to restrict production and adhere to the "stint" that limits the amount of work everyone will perform, the new economy appears to operate with the opposite pressure to "keep up" with everyone else (Burawoy 1979; Hochschild 1997; Shih 2004). One reason for this shift has been the introduction of managerial approaches modeled after those adopted in Japan. These systems create work teams, and are structured so that each member of the group becomes a supervisor for everyone else, creating a collective peer pressure to work harder (Parker and Slaughter 1988). This type of informal social control may be proving itself to be more powerful in extracting labor than the older managerial systems that relied on supervisors "driving" workers with a top-down system of control.

Another aspect of contemporary work that leads to intensified effort is the need to do many things at once. Jobs in the new economy require workers to "wear many hats" and to have the capacity to "multitask." Time management and prioritization have become mantras of contemporary human resources departments not just because workers have too little time, but also because they have so many (different) things to do. In addition, workers who can perform many tasks are less likely to find themselves with "down time" on their hands—an employee with nothing to do is likely to be reassigned to one of the several other tasks he or she is capable of performing. Compounding the problem further are the ways blurred work–family boundaries open avenues for workers to labor while at home (e.g., respond to e-mails) and to think about family while at work. As the anthropologist Charles Darrah (2005) observed, these time-strapped families do not "balance" work and family; their lives are endlessly busy.

Finally, workers in contemporary workplaces are often pressured to labor with intensity because employers deliberately choose to limit staffing levels to keep their companies "lean and mean." The result of these efforts to reengineer and downsize corporate workforces is a smaller number of workers carrying out more work. And in a context of pervasive job insecurity and weakened union power, individual workers feel intense pressure to work hard, lest the axe fall on their necks next. A poignant example is offered by American air traffic controllers, whose union was shattered during Ronald Reagan's presidency, and whose growing workload has been a subject of controversy ever since (Vaughan 2006).

Why Are Americans Working So Much?

A variety of factors contribute to Americans working longer hours than workers in most other countries, but chief among these is economic need. For those workers at the bottom of the opportunity divide, working long hours is a means of making ends meet. Consider the fact that the 2007 federal minimum wage rate of $5.15 per hour (now slated to increase to $7.25/hour) was the same as it was in 1996, and that its purchasing power was at its lowest level since the 1950s.[4] After adjusting for the effects of inflation, to achieve the same income earned by a full-time (40 hour) minimum wage worker in 1970, a similarly situated worker would have needed to work 55 hours in 2007. A full-time minimum wage weekly paycheck of $206 is barely sufficient to provide for individual needs, let alone enable a worker to raise a family, so some workers work multiple jobs or willingly work beyond the 40-hour norm. It is also important to recognize that many minimum wage workers are employed as "part-time workers" but work for multiple employers, thus putting in full-time (or often greater than full-time) commitments with no prospects for overtime pay.

Economic considerations also contribute to the overworking of individuals who hold managerial and professional jobs, but for these workers, career considerations or the threat of job loss is the motivating factor. Workers in such jobs fear that resisting pressures to work long hours will be interpreted as a lack of commitment, hurt their chances for promotion, or "mark" them as expendable when cuts are to be made (Fried 1998).

Beyond economic factors, American workplace cultures play an important role in pressuring workers to labor long hours. One of the most important studies to reveal this dynamic is Arlie Hochschild's (1997) *The Time Bind*. This book examined the work and family lives of professionals employed at "Amerco," a company that offered numerous family-friendly programs, including options for reduced hours and family leave. To her surprise, Hochschild observed that even when Amerco introduced family-responsive policies, few workers used them. This opens an interesting question—if some American workers have access to leaves of absence or reduced hours, why don't they take greater advantage of existing opportunities to work less?

The answer can be found in organizational cultures that encourage Americans to see long hours of work as evidence of personal competency (Blair-Loy 2003; Fried 1998; Shih 2004). Workers who occupy "good jobs" in the new economy labor with peers and supervisors who distribute badges of honor to those who can put in long hours and stigmatize those who do not. In companies like Amerco, workers who failed to put in long hours were viewed by their coworkers and supervisors as "nonplayers" who lack the qualities needed to move up the corporate ladder. The prospect of falling off the career track kept many working longer than they wanted, but others recognized that a choice to scale back could reduce their jobs to the least interesting and rewarding tasks (see Barnett and Gareis 2000).

The Time Bind also offered a radical new interpretation that inverted the "home is a haven in a heartless world" thesis, concluding that many workers in the new economy find their lives enhanced in the workplace, rather than in the family. One of the subjects in the study explained why she gladly puts in overtime:

> I walk in the door and the minute I turn the key in the lock my older daughter is there. Granted, she needs somebody to talk to about her day . . . The baby is still up. She should have been in bed two hours ago and that upsets me. The dishes are piled high in the sink. My daughter comes right up to the door and complains about anything her stepfather said or did, and she wants to talk about her job. My husband is in the other room hollering to my daughter "Tracy, I don't *ever* get any time to talk to your mother, because you're always monopolizing her time before I even get a chance!" They all come at me at once.
>
> Quoted in Hochschild (1997 p. 37)

Thus, in contrast to the romantic ideals of family life, *The Time Bind* argued that time-strapped Amerco workers experienced home lives of whining children, nagging spouses, and an unending backlog of onerous household chores. By contrast, their jobs offered friendly coworkers, concrete rewards, and interesting tasks, leading workers to *prefer* to be in the office than in the home.

Hochschild's study draws attention to two crucial aspects of contemporary American life: the difficulties of home life in families where both parents work, and the fact that many workplaces offer real rewards to workers willing to put in long hours. Nevertheless, Hochschild can be criticized for overstating the degree to which all American workers are *voluntarily* choosing work over home. Even some of her own interview subjects indicated that they worked long hours because they *had* to (Meiksins 1998). It is also important to remember that the type of worker studied at Amerco tended to be from the creative class, representing the values and perspectives of workers whose jobs offered numerous intrinsic and financial rewards (Florida 2002). The lives of these employees stand in stark contrast to the legions of workers, such as maids, fast-food workers, or retail clerks, who labor in backbreaking jobs or in hostile work environments. Even within its analysis of the creative class, the Amerco study failed to consider other professional workers who engage in creative work, but who are exploited by new organizational structures. For example, colleges and universities employ increasing numbers of part-time adjunct faculty, who commonly work long hours, often at multiple jobs, for a fraction of the pay commanded by their full-time counterparts.

Long hours in the workplace offer opportunities to earn badges of honor for American workers, but this cultural orientation to work is less common in Europe. Our colleague Peter Whalley observed that Americans take pride in broadcasting their heavy schedules to coworkers to increase their status. In contrast, when German workers put in long hours, their colleagues view them as inefficient and wonder what is slowing them down. Thus, cultural differences may account for some of the disparities between American workers and European workers. But there are structural reasons as well. Interestingly, although Americans have been increasing their work hours, European nations have instituted shorter workweeks. For example, French companies are legally discouraged from employing workers beyond 35 hours a week (Fagnani and Letablier 2004). This legislation reflects a desire to reduce the problem of overwork, to translate productivity gains into increased leisure, and to spread the available work around to all members of the society.

That European workers labor shorter hours than their American counterparts has led to some controversy. European employers (and some European governments) fear that overall economic productivity and competitiveness are being harmed. Voices have been heard in many European countries (notably in Germany and France) calling for a relaxation of limits on the workweek and a move to a more American-style approach to employment. Counterarguments and proposals have also been made, and there has been an effort to "standardize" work hours across Europe by establishing a ceiling for European Union countries on how many hours a worker can be required to work (at 48 hours). This would force some countries, such as Great Britain, to alter their current practices. The controversy continues and, so far, the gap between American and European work hours remains largely intact (Meiksins and Whalley 2004).

Another reason for overwork in the United States is that Americans face greater penalties than Europeans if they choose to reduce their work hours. For example, most working families in the United States secure medical care through an attachment to a full-time job. In contrast, citizens of many

European countries can cut back on work, or exit the labor force entirely, and still retain access to medical care (Meiksins and Whalley 2004). Similarly, most other developed nations have invested in extensive public transportation systems, infrastructures that afford residents the option not to purchase automobiles. In contrast, the minimal public transportation infrastructure in the United States builds in an expectation that Americans purchase cars and pay for their upkeep (Feagin and Parker 1990).[5] Consider, too, the impact of educational funding in the United States, where the system ties school quality to local property values. This system entrenches social inequalities and propels economic competition to purchase property in desired school districts (Kozol 2006; Marshall and Tucker 1992).

Finally, overwork is structured into the American system of allocating jobs. To understand how this happened, we need to go back to the Great Depression, an era in American history remembered as the time when many citizens lacked jobs. Less commonly recognized is that during the Great Depression, those who did have jobs commonly worked very long hours. To resolve this dilemma, the Roosevelt administration introduced overtime provisions as part of the Fair Labor Standards Act (FLSA). This law introduced a financial penalty for overworking employees, requiring employers to pay time-and-a-half for each additional hour worked beyond 40 hours per week. This legislation helped redistribute work opportunities from those laboring too many hours to those not working at all (Costa 2000). But because of the high costs of health insurance and other employer-sponsored benefits, it is now far less expensive to expect a smaller number of full-time employees to work overtime than to hire a larger number of full-time workers to work a 40-hour week (Tilly 1995). These expenses also increase incentives to create part-time or temporary jobs that are not eligible for benefits. When employers face pressures to increase production, they tend to do so by asking their current workforce to labor more or by hiring workers to whom they do not have to pay benefits.

The FLSA defined two classes of workers: hourly workers who are entitled to overtime pay, and "exempt" salaried workers (most commonly employed in managerial or professional jobs) whose work hours are not regulated. These exempt employees are among the most time-strapped in the new economy, having experienced an incremental expansion in job expectations during the past century. As Jacobs and Gerson (2004) note in *The Time Divide*, the exemption of some workers from overtime rights has increased the incentives to overwork salaried employees and to push more workers into jobs designed to have fewer protections. One illustration of this process can be found in the ways job boundaries can be blurred in the new economy. Such is the case for "team leaders" on production crews—workers who are primarily engaged in production work, but who also have some supervisory responsibilities. How should these employees be classified—as hourly workers with rights to overtime pay, or as salaried workers who have no such rights? In 2004, the Bush administration passed legislation that excluded some hourly workers who performed supervisory work, as well as those achieving higher incomes, from overtime eligibility. The Economic Policy Institute estimated that this simple act of redefinition, which redrew the boundaries of exemption from the FLSA, resulted in 8 million fewer workers being entitled to overtime pay. It also created incentives for employers to shift hourly workers into salaried "management" positions that, in reality, have few managerial responsibilities. What changed were not the day-to-day tasks, but rather, the right to extra compensation when laboring beyond the 40-hour threshold (Eisenbrey and Bernstein 2003).

In sum, why are Americans working so long and hard? Part of the answer rests in economic need—reasonable hours for those at the bottom do not provide a livable income, and reasonable hours for those in the fragile middle do not secure a stable career. Other answers can be found in an American culture that praises those who overwork, as well as policies that encourage overworking employees.

Nonstandard Schedules: Jobs in a 24/7 Economy

The old Dolly Parton song "Nine to Five"—the theme for the film of the same name—included the exasperated statement, "What a way to make a living." Although there are still many working women (and men) laboring the traditional 40-hour-a-week grind, they compose a far smaller proportion of the workforce. Workers in the new economy adhere to a diversity of schedules, including part time, split shifts, full time, and overtime. Fewer work by the hour and more are employed by the contract or in salaried positions that demand 50 to 70 hours of labor per week (or more). Workers labor during the day, but also in the evening, at night, and on the weekend. Although some clock in and out of their jobs, other employees are placed "on call" and can be summoned to work at a moment's notice. Although some workers have traditional schedules, more are demanding flexible arrangements. For still others, flexible schedules are designed by employers to meet their own needs.

The reorganization of work, allowing it to be performed around the clock, and just in time, is one of the contours of work in the new economy. The new economy offers no week "ends"; it continues 24 hours a day, 7 days a week. In the United States, there is no Sabbath, blue Sunday, or holiday that compels workers to take a break. Notice, for example, how some stores proudly announce that they are open "24 hours a day, 365 days a year." As the demand for workers to fill jobs around the clock increases, so too do the challenges faced by working families. The elaborate family schedules (typically posted in kitchens) that reconcile conflicting obligations of working parents and children attest to the complexities of melding work and family schedules.

Only a few decades ago, most stores were closed in the evenings and on Sundays, and most workers held down a regular 9-to-5 schedule. In contrast, the new economy operates on entirely different schedules. Just as the Industrial Revolution forced the culture to adjust to new rhythms, the emergence of a 24/7 economy is creating profound changes in work and family lives, as well as in other institutional arrangements. This presents new opportunities for working in new ways, but also new challenges in adjusting lives to nonstandard work schedules.

Exhibit 6.6 illustrates one means of capturing the extent to which nonstandard work schedules are proliferating in the new economy, revealing

Exhibit 6.6   Distribution of Work Schedules: All Workers, United States 1997

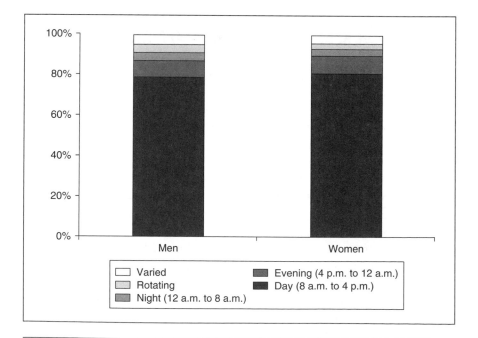

Source: Presser (2003)

Note: This exhibit represents all workers, including those who work full time, part time, and more than full time.

that one in five men and women is employed in jobs that involve working most of their hours in the evening, at night, or on rotating shifts that straddle conventional time divisions, or have schedules too complicated to be categorized. Note that these schedules represent data from all workers, including those who work part time, full time, or greater than full time. When the number of work hours is taken into account, *fewer than one in two employees (40%) works a standard day shift approximating the notion of a 9-to-5 full-time job.*

Part of the reason for the expansion of nonstandard scheduling is the emergence of new markets. For instance, evenings and weekends present lucrative opportunities for retailers to increase sales. The same is true for employers in the fast-food industry, which in the 1980s expanded business hours to enable the introduction of breakfast menus. The importance of offering shopping hours at "nonstandard" times has grown with the increasing prevalence of dual-earner couples. Few families can function if stores and other businesses are open only between 9 and 5, and many time-strapped families rely on restaurants and other commercial outlets for services they once performed themselves.

New managerial practices and philosophies also increase pressure on manufacturers to extend work into the nights and weekends, thus contributing to the creation of nonstandard schedules. Flexible production methods discourage employers from keeping a large stock of component parts ready for assembly, instead favoring the "just in time" approach, in which parts are manufactured (and often received) close to the time at which they are going to be needed. This new approach helped companies such as Harley-Davidson modify product lines with greater ease, minimize persistent quality problems, and create products tailored to individual customer desires (Reid 1989). But producing goods "just in time" requires the creation of demanding production schedules that may operate into the evenings and weekends. In the wake of expanded markets and reconfigured production methods, distribution companies such as UPS have expanded their operations, employing legions of workers to labor throughout the night processing packages, enabling a lobster caught in Nova Scotia to be transported live to a restaurant in Kansas (McPhee 2005).

Why do people work nonstandard shifts? Part of the answer is found in job markets and the types of opportunities available to workers. For those in the lower classes, where well-paid work is more difficult to find, the need to find any work can require bending one's schedule to fit the needs of employers. The most common reasons for working odd hours are that they are built into job definitions and that opportunities for working a regular shift are not available. The next most common reason offered—and this is more often the case for women—is that nonstandard job schedules offer greater opportunity to manage personal or family needs, such as helping with child care or enabling workers to continue their education (Presser 2003; Wharton 1994). It is not unusual for spouses to engage in a tag-team strategy of managing family—as one spouse returns from work to home, the other spouse leaves the home to go to work. But this adaptive strategy comes at a cost, as spouses' lives are adversely affected by their short contact with each other, and as the bonds that unite couples weaken in the absence of rituals such as the evening dinner (Ochs, Graesch, Mittmann, Bradbury, and Repetti 2006). An award-winning study revealed that recently married couples with young children who adopted alternating schedules were *six times more likely to get divorced* than were similar couples who worked regular schedules (Presser 2000).

How Americans Deal With Overwork

Having a good job is often accompanied by the expectation that one will be able to labor long hours, uninterrupted, year in and year out. Those who hold bad jobs, many of which are located in the expanding service sector, need to labor extremely long hours to afford even a scaled-down version of the American dream. Foregoing long hours comes at great costs, and many American workers face a stark choice: accept long hours as a way of life or choose to work shorter hours for considerably less compensation or fall off the career track. As a result, the most common response of workers is not to challenge employers or government leaders to change workplace policy but rather, to bend their lives to fit within existing structures (Moen and Roehling 2005).

Younger, childless workers are more apt to be willing to put in long hours, whereas those later in the life course, who have caretaking responsibilities, or whose health may be failing, experience greater difficulty adjusting their lives to fit the expectations of employers. Interestingly, most young men and women do not plan to make radical changes in their work hours following the birth of a child. But after trying to "have it all," many find the demands of work and family too overwhelming to manage. Faced with mounting pressure, the most common response is to fall back on gender roles developed in the old economy (Becker and Moen 1999).

One way of capturing this dynamic is by studying the different schedules dual-earner couples adopt before, and after, they have children. Exhibit 6.7 shows the work-hour arrangements adopted by two groups of dual-earner couples, those who are younger nonparents (dual incomes, no children), and those who have a young child. Five work-hour arrangements are identified:

Exhibit 6.7 Work-Hour Arrangements of American Middle-Class Dual-Earner Couples

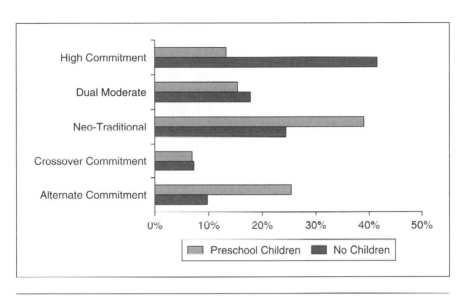

Source: Adapted from Moen and Sweet (2003)

- High-commitment couples (both spouses working long hours—45+ hours a week)
- Dual moderate couples (both working full time—35 to 45 hours a week)
- Neo-traditionalists (husband working long hours but the wife not)
- Crossover-commitment couples (wife working long hours but the husband not)
- Alternate-commitment couples (neither working long hours and one working reduced hours).

Notice that in the new economy, the high-commitment arrangement (with both partners working long hours) is the most commonly adopted schedule for young nonparents. In contrast, the most common strategy for couples who have preschool-age children is to adopt either a neo-traditional arrangement (the wife scaling back her work commitments) or for both partners to work less than long hours. This arrangement remains largely the same until children mature and leave the household (Moen and Sweet 2003).

Most Americans espouse (at least in words) fairly egalitarian attitudes regarding women's and men's opportunities, believing that all members of society should have the same options and equitable rewards. But Exhibit 6.7 reveals gendered responses to time strains. When faced with a deficit of time available for care work, families tend to revert to the patterns laid out in the 19th and 20th centuries, normative arrangements that push women into the home and men into the workplace. The result is the perpetuation of gendered opportunity structures, arrangements in which men have careers while sacrificing family time, and women sacrifice careers for the rewards of family.

These time strains are resonating in a variety of other ways. Nearly half (43%) of all workers feel that they do not get enough sleep, and most workers do not feel highly successful at "balancing" work and family (Hochschild and Machung 1989; Moen, Waismel-Manor, and Sweet 2003).[6] Faced with the choice of "making a career" or "making a family," Americans are now having fewer children, having children later in life, or sometimes even giving up on family altogether (Altucher and Williams 2003; Farnesworth-Riche 2006). Consider the startling fact that women now have their first child nearly 4 years later than they did half a century ago. According to the *Statistical Abstracts of the United States*, in 1980, only 1 in 14 (7%) women age 40 to 44 was childless, but by 2003, this figure had risen to nearly 1 in 5 women (19%). Families who do have children are increasingly obliged to make use of child care providers (professional caregivers, neighbors, relatives), even though they are often reluctant to do so. Job demands appear to be having a profound impact both on the shape of family lives and on the prospects of forming families in the first place.

Some workers are demanding changes, requesting that their employers rework schedules to accommodate personal needs. Technological innovations and the rise of service work have made it possible for many jobs to be performed in settings outside the workplace (the home, in transit) and at a variety of times. These changes open the prospects for **flextime** and **flexplace** schedules that enable workers to labor at home; to negotiate the number of hours expected; to refashion the starting and ending times of work as well as break lengths; and to create opportunities to bank time (Christensen and Staines 1990). Advocates of this approach to redesigning work point to the fact that it does not necessarily involve reducing how many hours jobs are expected to take, but does require expanding the amount of control workers have in setting schedules that match their needs off the job (Negrey 1993). Although not as effective in fostering work–family balance as many had hoped, evidence suggests that flexible work arrangements tend to create greater employee loyalty and to reduce turnover (Kossek, Lautsch, and Eaton 2006; Kossek and Lee 2005; Lambert 2003; Roehling, Roehling, and Moen 2001). Some workers, particularly those with scarce skills, have been able to arrange reduced work schedules without suffering the severe economic and career penalties usually linked to part-time work (Meiksins and Whalley 2002; Tilly 1995). Although flexibility offers new opportunities to bridge work and family more effectively, new work arrangements create new concerns for working families—especially in respect to establishing boundaries that enable one to disengage from work. As some workers have found, the option of working into the evenings or at home creates new portals through which work can intrude into leisure time and domestic life (Kossek and Ozeki 1999; Nippert-Eng 1996).

Conclusion

In the new economy, honor is bestowed on those employees whose cars are the first ones in the parking lots and the last ones out. Those who cannot keep up with escalating job demands, or meet the higher requirements to get jobs in the first place, are forced to the margins into jobs that offer insufficient rewards and opportunities. For the growing number of salaried workers, there is no cap on the number of hours that employers can demand or limits on the expectations of what it takes to keep a job.

Even though organizational and technological innovations have opened the possibilities to liberate work (such as through the use of flexible arrangements), they are being implemented in a culture that still equates productivity with time spent at work and that fails to prevent overwork. As a result, requests for shorter hours, or failures to put in "face time," usually come with career penalties. As American culture has embraced overwork as a virtue, the new economy has unleashed new sets of pressures (largely unchecked) to labor longer hours with intensified effort. This in turn has helped to create work poverty, limiting the availability of jobs for others who want them.

For workers at the bottom of the opportunity structures in the new economy, schedules and wages intertwine in problematic ways. Current minimum wage standards, for example, do not require employers to structure work schedules in a manner that enables employees to make a living; the standards require only that employers pay a (now woefully insufficient) base hourly wage. Even the more ambitious efforts to ensure a living wage (estimated at roughly $9/hour in most locales) assume that workers who receive this wage would have the ability to work full time. Little public discussion has been directed to the need to provide adequate compensation and supports to those who need to work less. Many workers—such as single parents—cannot maintain full-time schedules and perform their family responsibilities outside the workplace. Moreover, living wages are, in almost every instance, calculated on the basis of day-to-day needs, leaving long-term needs such as paying for college tuition or saving money for retirement unaddressed.

Resolving these issues requires tackling—head on—the institutionalized practices that push workers to labor long hours and countering cultural values that keep Americans from putting work in its place. In some respects, contemporary workers are facing as profound a transformation in their society as their counterparts did during the Industrial Revolution. Rethinking the concept of a fair day's work—how much, how hard, and how long employees should be expected to labor—needs to be front and center in a national dialogue on ways to refashion the new economy.

Notes

1. Before becoming an academic, one of the authors of this book was a professional carpenter and worked for a company that required him to haul tools from the equipment shed and make them operational before he could "clock in," and wrap up equipment after he "clocked out." This added more than a half an hour of uncompensated labor to each day.

2. Long vacations, however, can create their own problems. Such was the case when a heat wave enveloped Europe in 2003. Because the French customarily take holidays during the summer months, elderly citizens had scarce access to health services, resulting in high mortality.

3. When the CPS calculates average work hours, it excludes those who are not in the labor force, which can create an impression of more time spent pursuing paid labor than actually occurs. The estimates reported in Exhibit 6.5 include working-age adults who are out of the labor force and who are categorized as working 0 hours. This reduces the estimated number of hours worked, but enables us to reveal the remarkable changes in the average working hours created by women's entry into the paid labor force.

4. Some states had already increased the minimum wage beyond the federal level, and a few established standards above the *new* $7.25 rate; Washington set the highest minimum wage standard—$7.93/hour (Morgan and Morgan 2007). The new $7.25 federal standard will be phased in over 2 years and will still leave the real value of the minimum wage below the peak values reached in the late 1960s and early 1970s (Economic Policy Institute 2006, 2007).

5. One in three motor vehicles in the world is located in the United States, and the U.S. vehicle-to-person ratio is nearly double that of countries in Western Europe (Freund and Martin 2000)

6. Authors' analysis of the Ecology of Careers Study. Data represent a sample of middle-class, dual-earner couples.

7

Reshaping the Contours
of the New Economy

There is little evidence that the new economy is moving in a direction that will ensure that everyone has opportunities to engage in meaningful work, will earn a comfortable income, or will have the resources to construct satisfactory lives outside of work. On the contrary, employer practices and institutional arrangements continue to sustain—and sometimes even deepen—the chasms that separate workers from opportunity.

In this final chapter, we consider ways to reshape the new economy so that opportunities and resources are more equitably distributed. To do this, we first revisit the issue of opportunity divides to highlight the major concerns facing workers and discuss what would need to change to address those concerns. We then turn our attention to the process of social change and strategies to make workers' interests a priority in the ongoing development of the new economy and some obstacles that will be encountered along the way. Our hope is that strategic action, built through individual and collective efforts, can bring about needed changes. Finally, we conclude this book with a brief reflection on the prospects for liberated work in the new economy, work that would match the potential contributions—and needs—of a diverse workforce.

Opportunity Chasms

Class Chasms

One of the major problems in the new economy is that so many workers find it difficult or impossible to find satisfying and secure jobs that pay. There are both old and new aspects to this problem. Although workers across the class spectrum face problems of strained schedules, insecure jobs, and uncertain futures, these concerns are felt most strongly by those working in the numerous "McJobs" that emerged as a result of efforts to deskill work in the old economy. Today, legions of workers labor in jobs that offer low pay, few benefits, and few opportunities for growth. These workers can expect scant rewards from their jobs and slim prospects that diligent efforts will result in upward mobility. And because women and minorities tend to be funneled into these low-end jobs, this class divide contributes to gender and racial inequalities.

Although bad jobs continue to exist at the bottom of the American economy, there also have been efforts to chip away at the securities and leisure time available to those in the middle class. Insecurity extends to nearly all segments of the workforce, and most workers contend with strong prospects that their (or their spouse's) employer will close shop, lay them off, or restructure their jobs away. Their careers commonly require working long hours and affect their ability to form rewarding family lives. These workers' lives remain very different from those laboring at the bottom, but a feature of the new economy is that many "good" jobs have acquired features once associated with less desirable employment.

At the same time, the transition to the new economy witnessed the richest members of society becoming even wealthier and further separating their lives from the rest of society. In contemporary America, the significant gaps are not just between the middle class and the poor, but also between the affluent and everyone else. It is hard not to conclude that the good fortunes of those at the top of the class structure have come from the bad fortunes of those at the bottom and the increasingly fragile middle.

What can be done? One of the most pressing concerns is to address the economic problems faced by low-wage workers and to reshape the terms under which their jobs operate. There needs to be a national dialogue, and action, to rebalance the equation of a fair day's effort and a fair day's pay, including issues of compensation, scheduling, and security. But beyond those issues, resources and institutional arrangements will also need to be restructured in ways that enable workers to move from dead-end, low-skilled positions into jobs that offer greater rewards. This will require, among other things, designing and supporting educational opportunities that fit the structure of the new economy. And because the returns on work are so lopsidedly allocated, the lion's share of the financial burden of paying for these changes must be borne by the affluent.

Issues of job insecurity, unmanageable schedules, tenuous access to health care, limited (and sometimes nonexistent) vacations, summary dismissal, and a variety of other concerns cut across class lines. Creating entitlements to reasonable treatment and reinforced safety nets will help all workers, not just those laboring in bad jobs. Launching initiatives to address these concerns will require abandoning cultural mindsets that divide groups of workers from one another. As long as low-wage workers see themselves as pitted against those with education, or the middle class sees itself as being pitted against those receiving welfare, the effort to mobilize support for collective resources will be hampered. As was the case in bringing about the Fair Labor Standards Act and Social Security legislation in the wake of the Great Depression, catalyzing change will almost necessarily require recruiting the fragile middle class into the struggle.

Gender Chasms

In the old economy, social policies, organizational cultures, job designs, and personal expectations were shaped by the assumption that men would be breadwinners and women would be homemakers. Within this gender regime, women's efforts in the home were defined as something other than real work, and care work was not given its economic due. In the new economy, women are nearly as likely as men to work outside of the home and to aspire to meaningful careers. And it is not simply women's desires that influenced their increased integration into the paid labor force—most families need two earners to make ends meet. Inequalities between men and women persist in no small degree because of the persistence of ideas and structures that place most of the burden of care work on women and impede their access to the most financially rewarding jobs.

Gendered work standards, modeled on what men used to be able to bring to the workplace, are now structured into job designs and employers' definitions of who are "ideal workers" (Williams 2000). Workers have been socialized to accept a career mystique, an unsustainable set of expectations promoting an intense commitment to work (Moen and Roehling 2005). These structures and beliefs chafe against the resources available to the new workforce, one composed increasingly of dual-earner and single-parent families. How are workers responding? Most commonly by laboring long hours, shifting the timing of life events (such as marriage and childbirth), doing without, and shouldering the burden of increased stress. Their ability to manage also is often predicated on their ability to employ low-wage workers to care for children or aging parents.

Recognizing these facts highlights the need to dismantle a gender regime that continues to assign men and women to different jobs inside and outside the home. But beyond countering the forces of socialization and interpersonal discrimination, structural changes will have to occur as well. Compensation will need to be recalibrated so that care work receives equitable returns as a form of labor. And career and job templates that assume workers have the capacity to work like men with full-time, stay-at-home spouses will need to be reconfigured.

Workers and employers in the new economy need to develop new definitions of how much work employees should be expected to perform and to redesign jobs to be reasonably compatible with what workers can provide across their careers. Efforts to create a "one-size-fits-all" approach to work and family strains are unlikely to succeed, as the potential contributions of most workers vary over the life course. A more fruitful direction for change involves fostering new templates for flexible careers (Moen and Sweet 2004). These can include opportunities to scale back work hours or to take time-outs from the labor force. And individuals and couples will need to have the resources to be able to effectively plan careers so that they can land on their feet when jobs are lost.

The issues surrounding gender and work are commonly cast as "women's problems," but the reality is that these are family problems that affect men, women, their children, and their aging parents. Accelerating the pace and extending the range of reform will require bringing men into the movement to humanize work by persuading them of the benefits to be reaped from reconfigured gender roles and resources. This is actually an easy argument to make. Which stressed worker would not want to see the establishment of reasonable work schedules, expanded vacations, and opportunities to secure time-outs from his job? If more men are convinced that they lose out with gendered divisions of labor—particularly in their access to time to spend with their children—the clamor for change will increase, and coalitions for change will strengthen.

Racial and Ethnic Chasms

Although Civil Rights–era legislation imposed legal barriers against discrimination, racial and ethnic ties continue to influence access to jobs. Today, race and ethnicity play a role in determining where one lives, the resources families have to pass from one generation to the next, and the attitudes gatekeepers hold about individuals from different backgrounds. Like the gender chasm, the race chasm is maintained through social structures that differentially allocate resources and opportunities, as well as cultural orientations that shape expectations.

If the new economy is to become truly race-blind, it will need to address—but also go beyond—sources of interpersonal discrimination, such as the widespread use of stereotypes that label potential employees as suspect. The disadvantages minority group members face stem from being raised in families that lack the economic resources to afford college tuition, the social connections to facilitate entry into good jobs, and the cultural capital that provides the soft skills needed for success in the new economy. But more than family disadvantages and interpersonal discrimination contributes to racial inequities. They also result from community-level resource deficits. The forces that concentrate members of underprivileged racial and ethnic minority groups into poor neighborhoods make it likely that they will be deprived of quality education, information about work and careers, and the types of social ties that link them to jobs.

Racial tensions continue to divide the U.S. workforce, with controversies over illegal immigration, affirmative action programs, and welfare reform all exhibiting strong racial overtones. The dialogue about whether illegal immigrants are "taking jobs from American workers" or "doing jobs that no one else will perform" distracts from the larger question of how to regulate low-wage work. A crucial step in reorienting this debate will be to shift the focus away from blaming the victims of globalization and economic restructuring, and toward analyzing the conditions of work—at home and abroad. It is interesting to note that virtually absent from the illegal immigration and welfare reform debates are discussions of the exploitative conditions under which most low-wage work occurs or of the role of employers who provide jobs to undocumented workers. Building racial coalitions is going to be a great challenge because prejudice and discrimination are etched so deeply into the culture and structure of American society. But these barriers can be dismantled by groups that share common interests (which we discuss later in this chapter), and as this happens, prospects for opening opportunities for underrepresented minority groups will be enhanced.

International Chasms

The new economy's global reach is profoundly affecting work opportunities around the world. It is impossible to ignore the remarkable transformations occurring in China, India, and many other developing nations.

Longer life expectancies, as well as access to medical care, educational opportunities, and consumer goods, reflect a variety of positive outcomes of globalization. But at the same time, globalization has contributed to a number of new problems in the developing world, including pollution, over-population of cities, and the creation of hazardous work environments. Although some societies are advancing, other places in the global economy—especially in sub-Saharan Africa—have been largely left behind.

The flow of work opportunities into the developing world and the movement of employment away from its traditional locations in the United States and Europe have contributed to American workers' feelings of inse-curity and fueled international antagonisms. Some American workers express hostility toward lower-priced workers overseas, who are perceived as stealing Americans' jobs. At the same time, the United States is increas-ingly viewed negatively overseas, not only because of recent political deba-cles, but also because its power is perceived as coming from the exploitation of workers abroad. The cheap consumer goods available in American stores often come at a high cost to workers in foreign lands—and these facts are well recognized within the developing world. As a result, the globalization of work may be catalyzing radical responses to what is perceived as American cultural, political, and economic domination. And even in coun-tries that are making economic strides, vast numbers of people are unable to achieve a standard of living even remotely like that enjoyed by an aver-age American.

The solutions to these problems are not likely to come from the free mar-ket. As Karl Marx observed, the forces that make capitalism so productive are intertwined with the drive to exploit labor, and to do so around the globe:

> The need of a constantly expanding market for its products chases the bour-geoisie over the whole surface of the globe. It must nestle everywhere, settle everywhere, establish connections everywhere . . . All old-established industries have been destroyed or are daily being destroyed. They are dislodged by new industries, whose introduction becomes a life and death question for all civilized nations, by industries that no longer work up indigenous raw material, but raw material drawn from the remotest zones; industries whose products are con-sumed, not only at home, but in every quarter of the globe. In place of the old wants, satisfied by the production of the country, we find new wants, requiring for their satisfaction the products of distant lands and climes. (Marx and Engels 1972 [1848])

The current system enables unfettered mobility of capital and creates incentives for employers to move jobs to locations where labor and the environment can be most easily exploited. Developing societies will not necessarily benefit from this movement because the same mechanisms that enable capital to move in also enable shifting the profits to other locations. Noted analysts such as Thomas Friedman (2005) optimistically assert that globalization will "flatten" the divides between developed and developing countries. Although this may be true in cultural terms, in economic terms the divides between developed and many developing countries are expanding (United Nations 2005). This is not to say that life will not continue to improve in most developing nations, but it will not be nearly as good as it could be if reasonable checks and controls, designed to ensure more equitable returns from work, were placed on free-market capitalism. This needs to be done not only for ethical reasons, but also because it is in the interests of workers in the United States to do so (Kochan 2005).

The Agents of Change

Throughout this book, we posed questions concerning the contours of the new economy and the origins of the divides that separate workers from opportunity. In this final section, we consider two remaining important issues—identifying the types of changes that are feasible and the forces that can make these transformations happen. Our argument is that refashioning the new economy will require the concerted efforts of multiple agents, ranging from individual consumers to activist groups, unions, employers, government, and international organizations. However, not every agent will carry the same level of force, and when their efforts are misapplied (or misrepresented), they can actually impede change rather than accelerate it.

The Role of Individuals

Efforts to reshape the workings of the new economy, and to remedy its problems, often encourage a "do-it-yourself" approach. But can the cumulative efforts of disconnected individuals lead to the types of changes needed to substantially refashion the contours of work and opportunity? Consider, for example, the advice manual *The Better World Handbook: From Good Intentions to Everyday Actions* (Jones, Haenfler, Johnson, and Klocke 2002). This book (like a number of others) suggests that the path to reform leads through self-reflection and conscious efforts to "do the right thing." Toward that end, consumers are advised to limit their trade with companies that are less than environmentally sensitive or labor friendly, as well as to treat coworkers and subordinates with the dignity that they deserve. Investors are encouraged to buy stocks and invest in companies that have a positive track record on issues of public concern. And, like innumerable other self-help books, it offers advice on the best ways to "balance" work and family.

Though intuitively appealing, these "do-it-yourself" solutions are unlikely to create truly meaningful change by themselves. One problem with the focus on the isolated consumer is that individuals face the Herculean challenge of identifying those brands and companies that are "responsible." To do so requires not only understanding the actions of distributors, but also the various companies that are linked to them in vast global supply chains. And companies can have mixed histories—for example, they may be strong on labor concerns but much weaker on environmental concerns. Socially responsible consumers also experience considerable difficulty determining which products to buy because the information they receive is often confusing and misleading (Nestle 2003; Seidman 2007). And even if consumers did have access to the relevant information, options to purchase responsibly do not always exist. This is a classic example of the exit/voice dilemma—individual consumers may possibly influence policy through exit (by avoiding choices they reject) but lack voice; that is, they cannot *create* options that correspond to their preferences. Of equal concern is the observation that consumers are reluctant to purchase fair trade products that cost much more than standard goods. In practice, people tend to place higher premiums on issues of style and price than on ethical business practices (Iwanow, McEachern, and Jeffrey 2005; Pelsmacker, Driesen, and Rayp 2005).

The expectation that individuals will provide solutions for collective problems has a long history in American culture. For conservatives, individualism reflects an American virtue, and they reference Tocqueville (1969 [1836]) in their observations of the ways their country's successes hinged on Americans' embrace of autonomy and volunteerism. Conservatives argue the merits of an "ownership society" that allocates as many resources as possible to individuals. Thus, rather than invest in a national health care system or Social Security, the individualistic approach is to offer health "savings accounts," and personally managed retirement portfolios. Nor is it only conservatives who place individualistic perspectives at the forefront of social policy; so do many liberals. This can be witnessed, for example, in President Bill Clinton's efforts to end "welfare as a way of life" and to introduce strict limits on support through the Personal Responsibility and Work Opportunity Reconciliation Act. In all of these cases, reforms were presented as empowerment, but actually had the consequence of shifting the responsibility for managing risk to individuals. As Jared Bernstein (2006) argues, these programs rest on an underlying philosophy that says "you're on your own" and that individuals have no right to expect support or protection from hardship (see also Hacker 2006).

Numerous cultural critics have argued that individualism results in a diminished capacity to empathize and understand how one's fortunes (or misfortunes) are connected to larger social processes (Bellah, Madsen, Sullivan, Swidler, and Tipton 1985; Mills 1959; Putnam 2000; Reisman 2001 [1961]). Even when sympathy exists, notice how individual volunteers are assigned the task of ameliorating hardship, supplanting bolder initiatives to challenge oppressive structural arrangements. For example, consider two of the most prominent volunteer and semi-volunteer groups in America today—Habitat for Humanity and Teach for America. Habitat for Humanity International, with a nearly all-volunteer membership, has erected approximately 200,000 homes in America and abroad since its inception. In 2005, Teach for America had 3,500 dedicated young teachers working in impoverished school districts. Without disparaging these accomplishments, it is important to place them in context. In 2006, in the United States alone, nearly 8 *million* families lived in poverty and *1 in 10* children did not graduate from high school. As well-intentioned as volunteers are, they commonly do little (if anything) to eliminate structural barriers or establish a more equitable distribution of social resources.

In sum, individualistic solutions are hampered by three fundamental concerns. First, people may not have the inclination or the capacity to behave in a manner that will transform the new economy. Second, individualistic efforts generally leave untouched the underlying forces that shape the contours of work and opportunity. Third, as a cultural framework, individualism reinforces the shift of risks to individuals, rather than building on the strength of collectivities. This does not mean that individuals cannot make a difference by trying to be socially conscious or that these efforts cannot create good outcomes. But these efforts alone will not be sufficient to address the root sources of the problems or make a significant impact on the operations of the new economy (Bernstein 2006).

The Role of Activist Groups

If individuals are not going to be the answer, how about activist groups? These groups comprise individuals who band together to exert pressure on governments, employers, and consumers. Their goals are to use collective action, primarily directed at the local level, to influence employment and trade practices, not only at home in America, but also abroad in developing countries. Here, we see greater cause for optimism, but the impact of their efforts will necessarily depend on recruitment, their alliances with other organizations, and their ability to identify and target institutions that can effect change.

Consider the successes of activist groups, such as United Students Against Sweatshops and United Students for Fair Trade, which have influenced their own educational institutions to sever contracts with producers with histories of labor abuse (Crawford 2003; Glover 2003). Among their accomplishments is increased public awareness of exploitative labor practices and their links to the production of well-known consumer items, such as those endorsed by P. Diddy, Kathie Lee Gifford, and Michael Jordan. By putting pressure on institutions from within, and by mobilizing public opinion, these groups engage in a "name and shame" strategy to persuade employers to change their business practices (Seidman 2007).

Other grassroots organizations pressure local governments and employers to implement reasonable terms of employment. For example, activist groups such as the Association of Community Organizations for Reform Now (ACORN) and Jobs with Justice operate almost entirely through local volunteers. Both of these groups organize social awareness campaigns and propose legislation at the local level to control employment practices (O'Brien and Gupta 2005). As a result of their efforts, more than 100 cities and jurisdictions had implemented living wage ordinances by 2003 (Freeman 2005).[1]

Although these successes provide reason for hope, it is important to assess them relative to the magnitude of the problems. Even the most successful grassroots initiatives tend to have limited reach in sparking change in employment practices in the new economy. In part, they are hampered by limited resources and fluctuating memberships. When they do effect change, gains can be fleeting, as companies remain free to move facilities to "employer-friendly" locations (Armbruster-Sandoval 2005). Another problem is related to the strategy of focusing change at the local level, which can sometimes be trumped by more macrolevel initiatives. For example, companies can influence the creation of legislation at the state level to restrict the right of local communities to set their own labor standards. And focusing on the local level tends to produce changes that are confined to small groups of workers in particular locations. Careful study of living wage initiatives, for example, reveals that comparatively few low-wage workers have been affected (most studies estimate the number to be less than 100,000, a small fraction of today's workforce) (Freeman 2005).

Arguably, the biggest success of activist groups has not been in local reforms, but in increasing public awareness of labor and environmental abuses. In response, companies have become more conscious of their public images, which in turn has influenced some major employers to reform labor

Exhibit 7.1   Student Protesters Outside a Gap Store in Manhattan

Source: © Najlah Feanny/CORBIS SABA. Reprinted with permission.

and trade practices. For example, once the widespread use of child labor gained public recognition, companies such as Nike, Reebok, and Adidas agreed to monitor their suppliers' employment practices. Coffee distributors such as Starbucks have signed agreements with "preferred suppliers," who have promised to pay better wages, not to employ child labor, and to have their operations monitored (Schrage 2004). Still, critics, such as the Organic Consumers Association, claim that Starbucks makes only limited use of "fair trade" coffee beans and that the company's commitment to economic justice is superficial. As we discuss later, employers have developed a variety of strategies to maintain the impression of being socially responsible, while often avoiding being truly responsible (Seidman 2007). The question that we will return to concerns the most strategic means for harnessing the collective power of these grassroots initiatives to force change.

The Role of Organized Labor

The American union movement has been in decline for much of the past 50 years and efforts to reverse this trend have proven largely unsuccessful. As we noted in Chapter 2, unions face a variety of problems in the new economy, including the decline of manufacturing employment, a legal framework that makes organizing difficult, capital mobility, the proliferation of smaller workplaces and subcontractors, and many more. These problems facing organized labor are compounded by public relations concerns. Many American workers have ambivalent feelings toward unions, which are commonly perceived as the reason why jobs move overseas, as protecting deadbeat employees, as corrupt, and as undermining the principle that individual workers (rather than classes of workers) should be compensated according to their efforts (Lichtenstein 2002). Only one in five Americans (20%) thinks that unions are "excellent" or "very good" for the country, one in three (36%) believes that unions block economic progress, and two in three (69%) think unions have enough or too much power.[2]

Although unions have diminished power in the new economy, it is important to recognize that they still hold considerable sway. Unions play a vital role in influencing job contracts for those who are members or who work in unionized workplaces. Today, even though smaller proportions of the labor force are organized, they remain a substantial political force, particularly within the Democratic Party (Dark 2001). Fully 1 in 10 workers is a union member, and many more Americans belong to a family in which at least one person is a union member.

There are some indications that unions *could* make a comeback in the United States. Despite ambivalent feelings about unions and their role in society, an estimated 40 million unorganized workers would vote for a union in their workplace if they were to be presented with the opportunity (Clawson 2003). Although the structure of the new economy and laws regulating union formation make organizing difficult, organized labor does have new weapons at its disposal. For example, "just-in-time" production techniques leave little margin for error if supply chains are to be effectively coordinated. This creates the possibility for strategically managed strikes to disrupt an entire company's operations (Parker and Slaughter 1988). New patterns of labor migration have also brought to the United States new groups of workers. Some of these recent immigrants are difficult to organize: Some are hostile to unions; others may have been frightened by labor repression at home; some are undocumented and fear deportation; and still others are "sojourners" and intend to return to their home countries. However, as was true in the 19th and 20th centuries, many immigrants (particularly those from Latin America) bring with them receptive attitudes to unionization and, in some cases, experience with organizing (Gordon 2005; Milkman 2006; Waldinger and Der-Martirosian 2000).

Several recent examples of successful organizing campaigns give proponents hope that the resurgence of union activity has already begun. In some cases, conventional unions have succeeded in organizing previously unorganized groups of workers. The Justice for Janitors campaign has made remarkable progress in organizing and winning better conditions for building maintenance workers (Milkman and Voss 2004). Similarly, the Hotel and Restaurant Workers Union succeeded in negotiating labor contracts with most of the full-service hotels in San Francisco (Wells 2000). Unions are even attracting the interest of professional workers, such as graduate students, who successfully gained union representation at several major universities (Lafer 2003). Surprisingly, some of the most successful organizing campaigns have been conducted by some of the oldest (and allegedly most "conservative") unions, those once affiliated with the American Federation of Labor (AFL). Ruth Milkman (2006) argues that this is because they are concentrated in growing sectors of the economy, in industries unaffected by capital mobility (such as hotels), and because their approaches are better suited to a volatile economy.[3]

Although conventional unions have had some successes, organizing low-wage workers, especially immigrants, sometimes requires another approach. One strategy that has shown promise is the creation of **worker centers**. These are community-based organizations, typically representing immigrants, that attempt to help low-wage workers in a variety of ways: providing services such as legal help and education; advocacy; and helping to organize workers (Fine 2006). Some have succeeded in improving wages, stimulating organization, and making legislative gains for workers, even in places like suburban Long Island, where low-wage workers are often undocumented and isolated from one another in small workplaces (Gordon 2005).

Several future directions for the union movement have been proposed. One view argues that unions must adjust to changing economies and production practices (Heckscher 1988). In a system where jobs are constantly in flux, and where manufacturing only constitutes a small portion of the job market, the old-economy approach of using seniority systems to protect workers makes little sense. New methods for protecting workers are needed, some of which may be learned from earlier efforts. For example, rather than employers controlling who gets jobs, unions could collectively manage the placement of employees as opportunities expand and contract. This method of managing insecurity was commonly used in the craft union halls of the early 20th century and its time may have come again (Lichtenstein 2002).

Others suggest that the future of unions depends on their regaining their militancy and energy. Only in this way, they argue, can unions overcome the negative image they have acquired and persuade workers that they are not bureaucratic, corrupt, and irrelevant (Lopez 2004). Dan Clawson

(2003), for example, calls for a new "upsurge" of union activity, which would require unions to organize new workplaces and industries, and, most importantly, change the political and cultural climate in which unions exist. He argues that something like this happened in the 1930s, when the last big flurry of new organizing activity took place. He believes this can happen again, if unions are creative and respond to new problems and opportunities.[4]

A resurgent union movement will have to confront globalization and develop a means of collaborating and organizing workers on an international level. This would not be an altogether new idea. Even in the late 19th century, labor "internationals" existed, and many 20th century unions belonged to a variety of international union federations. Most recently, an attempt to consolidate these international organizations led to the formation in 2006 of the International Trade Union Confederation (ITUC), which claims more than 300 affiliates in a wide range of countries. However, most international labor organizations, including the ITUC, do not engage in organizing efforts, but rather serve as clearinghouses for information, protest various forms of injustice against workers around the world, and pressure national and international organizations (such as the World Bank) on behalf of workers. Today, there are other opportunities for cross-national cooperation, such as an American union working together with a sister union at a branch plant overseas. Alternately, workers in one country could be organized to act in sympathy with workers in the same industry overseas. This happened in 1998, when American workers refused to unload ships affected by an Australian dock strike.

Unions will be better positioned to promote change when they integrate themselves with other activist groups, such as the anti-sweatshop campaigns and living wage efforts mentioned earlier. In the past, unions and new social movements have not been receptive to one another. During the Vietnam War era, unions were often perceived (sometimes accurately) as opposed to the student movement. Nevertheless, when unions work in concert with other grassroots organizations, an approach that has been called "social movement unionism," the results are often superior to when they operate alone (Gordon 2005; Milkman 2006). Successful living wage campaigns illustrate this well; so do the much-discussed demonstrations in Seattle in 2001, in which labor groups cooperated with church groups, environmentalists, and others to protest the meetings of the World Trade Organization and its policies regarding global trade (Clawson 2003). And the remarkably successful organizing drives in Las Vegas hotels and casinos also owe much to a conscious strategy of working together with other community organizations (Fantasia and Voss 2004).

In short, labor organization could, again, become a powerful force for workplace reform. However, there is no guarantee that this will happen. Obstacles to union organization remain powerful and the gains discussed here are fragile. The New York University (NYU) graduate students' union, which organized in 2002 and made substantial gains in wages, benefits, and rights to overtime, was essentially shattered in 2005 (Arenson 2005; Epstein 2005). Worker centers, too, often find their gains to be temporary or unenforceable and struggle both to forge links with nonimmigrant groups and to grow beyond a few hundred members (Gordon 2005). So, although it may be premature to announce the death of American unions, unions are unlikely suddenly to become powerful enough to reshape American workplaces by themselves.

The Role of Employers

The tensions between work and family life, and the growing awareness of the problems of global trade, are creating pressures to implement more socially responsible designs for workplace operations. But are these pressures sufficient to motivate employers to advance meaningful change from within? What is the likelihood that "internal" actions by the leading agents in the new economy will foster new approaches to work that serve the interests of workers and the communities in which they live?

One strategy to make employers take ownership of change relies on demonstrating how responsive policies can actually benefit companies. Work–family scholars call this the "dual agenda," and posit that what is good for working families, or the wider society, does not necessarily come at a cost to employers. Some family-responsive policies, such as flexible work arrangements (when applied in specific types of settings to specific types of workers) can actually work in the employer's interests (Pavalko and Henderson 2006). Fulfilling the dual agenda requires demonstrating the potential for positive "returns on investment" that can be measured by increased productivity, greater profits, better employee retention, consumer loyalty, or other indicators of success. This **business case** for reform shows employers how their companies can benefit from altering production practices and taken-for-granted ways of working (Kossek and Fried 2006).

In many industries, a business case can be made to enhance labor standards through increased wages, benefits, or the creation of flexible work arrangements. For instance, recognition as a "best employer" in publications such as *Working Mother* magazine can help a company market itself to consumers and potential employees. Similarly, a business case can be made to respond to brand tarnishing that results from "naming and shaming" pressures brought to bear by activist groups. A marketing business case can be made for developing products that fill the emerging niche for "fair trade" products. And opening opportunities for bridge jobs or flexible employment can provide a means to retain the talents, experience, and knowledge of an aging workforce (DeLong 2004). Researchers at the Massachusetts Institute of Technology (MIT) have worked with employers to design collaborative experiments to test how dismantling and replacing entrenched ways of organizing work affect the bottom line (Bailyn, Bookman, Harrington, and Kochan 2005). Other researchers, in collaboration with employers, are documenting the extent to which companies are implementing worker-responsive practices and then publicizing findings from which companies can "benchmark" their own performance relative to that of the wider corporate community (Harrington and James 2006).

Although these approaches show some promise, the employers in these studies tend to be those who want to retain and focus the energies of skilled workers. It is much harder, and sometimes impossible, to identify a business case to improve the conditions of work for those engaged in low-end jobs at home or abroad. If one traces the history of work from industrialization forward, it becomes apparent that the business case is often strongest for *dismantling,* rather than implementing, worker-responsive practices (such as providing high wages, secure jobs, health insurance, retirement benefits, and safe working environments—all of which have been eroded in the new economy). This is not to say that a dual agenda cannot sometimes be satisfied, especially for flexible work arrangements for highly skilled workers. But it is important to recognize the limitations of this approach because it is unlikely to close many of the major opportunity chasms, such as those that channel the lowest strata of the workforce into poorly paid, low-skilled work. For those workers, the business case (established by history and numerous current cases) is to degrade work and opportunity.

Beyond the business case strategy, others suggest that employers may move the new economy forward as they embrace an ethos of **corporate citizenship** or **corporate social responsibility (CSR).** In this case, the internal force is an open commitment, by corporate leadership, to understand their roles as stewards of the new economy (Blowfield 2005). CSR requires managers not to direct their actions simply to expand profit within the narrow confines of the law, but also to follow normative standards of professional conduct that go beyond what is obligatory or customary (Carter 2004). Perhaps most familiar to American consumers is Ben & Jerry's ice cream, a company that maintains that community-mindedness is a core component of the brand.[5] Beyond this company, the adoption of corporate social responsibility codes has been demonstrated to have a positive impact in the textile, leather tanning, and shoe industries in South Asia and in the wine industry in South Africa (Luken and Stares 2005; Nelson, Martin, and Ewert 2005; Winstanley, Clark, and Leeson 2002). And individual socially responsive leaders can make a difference in the lives of their workers. For example, because of the challenges of being a single parent following his wife's death, CEO Lewis Platt created sweeping changes at Hewlett-Packard, including the introduction of expansive flextime, flexplace, and family leave policies (Abelson 1999).

Will grassroots pressures and CSR lead to a more humane new economy? Gay Seidman (2007) offers a skeptical appraisal, basing her conclusions on the ways companies reformed (or failed to reform) their labor practices in South Africa, India, and Guatemala. She found that, even when the need for change was the focus of considerable public attention, corporate leaders largely failed to demonstrate a commitment to social responsibility. For example, even as the South African system of apartheid was exposed to widespread condemnation, American-based multinational corporations avoided challenging the South African government on human rights violations. Their primary goal was to keep operating within an oppressive regime. And when corporations did respond, they were slow, tempered, and acted in ways that respected the political system enforcing differential treatment of white and black South Africans. Even when companies signed pacts that required fair labor treatment, they were lax in enforcement, swayed evaluations to oversell their successes, and underreported the incidence and the implications of their failures.

Seidman concluded that companies, on their own, will not "do the right thing." Nor will pressure directly applied by consumers or activist organizations necessarily create a market-driven response that will lead corporations to "come to their senses" and act in responsible ways. A case in point is the much-publicized Rugmark program in India, a voluntary program by manufacturers to not employ children in the manufacture of carpets. In return, participating employers receive a Rugmark tag that is sewn to the back of the carpet. This is then used as a marketing tool, giving consumers comfort in believing that the rug they are purchasing is made in accordance with fair labor practices. In sum, the Rugmark approach relies on a voluntary program that employers sign on to, which in turn is used to ratchet up the value of their product (and puts pressure on other employers to behave in a similar socially responsible fashion).

Although Rugmark was initially lauded as an innovative market-based approach to stamping out child labor, subsequent analysis of its effects revealed serious flaws. One problem is lax oversight because the program is not supervised by the Indian government or international agents, but by a small number of inspectors employed by the Rugmark program. Inspections are rare, so that even when a rug is certified, it is uncertain that children were not involved. Rugmark also spawned competing marks, such as "Care and Fair," that assure consumers that a portion of the proceeds from their purchase will be returned to benefit children and others in the carpet district. This latter approach leaves intact the practice of employing children in grueling, hazardous work, and leaves consumers with the false impression that their purchase is child labor–free. Another concern is that the voluntary approach to enforcing fair labor practices can foster a two-tiered distribution system, in which some products produced by the same manufacturer can be exported to societies (or stores) that care about fair labor practices, whereas others can be exported to societies that do not. Finally, as Seidman (2007) notes, fair trade labels may matter for luxury or visible product lines such as rugs, but they cannot be effectively applied to other products (such as ball bearings) that are hidden from the consumer.

Is corporate leadership likely to guide companies to be "good citizens" in the new economy? Or are they more likely to continue to engage in "a race to the bottom" and seek ways to maximize profit at the expense of workers and the communities in which they live? As we have suggested throughout this book, considerable evidence suggests that the latter is more likely to occur. It is important to remember that capitalism itself discourages social responsibility because there are clear financial motivations to employ populations that can be most easily exploited and to move to locations where environmental protections are weakest. The case of Wal-Mart, the world's largest employer and retailer, amply demonstrates this proposition. In the wake of widespread condemnation of its employment and trade practices, Wal-Mart has directed far greater attention to its strategies for managing public opinion than it has to reforming its labor practices, and its profits and power continue to increase (Goldberg 2007).

The Role of Government

Political conservatives commonly caution against the interference of "big government" and the impediments regulation creates for the functioning of the economy. This perspective on government's role in the regulation of the new economy has a number of problems. First, government is *always* involved in economic activity. As the economic historian Karl Polanyi (1944) argued, the market itself could not exist unless the government had been there to help create it and continued to provide the political, legal, and military framework within which it operates. Second, government has a long history of making trade possible, and of shielding workers from abusive work conditions. Most of the protections workers have today are the result of government intervention; these include child labor laws, unemployment insurance, environmental regulations, social security, minimum wage standards, rights to overtime pay, workplace safety oversight, the right to unionize, prohibitions against discrimination, and the right to family medical leave (see Appendix A for a regulatory timeline). Those who espouse laissez-faire capitalism often overlook these necessary "intrusions" into the economy (Bernstein 2006; Galbraith 1976).

In the preceding chapters, we argued that the U.S. government generally has been reluctant to interfere in the contests between workers and their employers regarding pay, schedules, security, and other conditions of work. And U.S. government efforts pale in comparison with the more active role that governments have played in Western Europe. We also observed that, during the 20th century, the U.S. government went from being virtually absent from the regulation of work to being much more involved. The directions government actions will take in the future, and its speed of response, are not foregone conclusions. As in the old economy, competing interest groups will sway legislation, and the process will be political, not inevitable. Ultimately, government action will hinge on the ability of individuals, activist groups, unions, and others to influence its operations.

What role should the American government play in the new economy? The types of action we advocate will require reenvisioning the regulation of work, in much the same way as happened during the New Deal. This reenvisioning should consider specific workplace regulations as well as the collective responsibility to provide for the basic needs of citizens, irrespective of their attachment to the labor force. We suggest three types of policy initiatives. First, there must be a revisiting of the issue of what constitutes reasonable conditions for work, and a revision of the standards enacted under the Fair Labor Standards Act (and other legislation) to consider both the nature of jobs and the composition of the labor force in the new economy. Second, there needs to be a national discussion of the means by which affluence and opportunities can be more equitably distributed. Third, although much attention has been focused on the U.S. government's role in promoting the interests of employers and protecting jobs at home, there needs to be a forceful discussion of its role in regulating the terms under which global supply chains operate and the role of the United States in fostering *positive* development abroad.

A Fair Labor Standards Act for the New Economy

When the Fair Labor Standards Act was enacted in 1938, it offered Americans, for the first time, the right to overtime pay and a minimum wage. Since its passage, its provisions have received modest updates and some new worker protections have been instituted. However, the regulation of work today largely fails to address the needs of many workers in the new economy. Today, employers can expect workers to labor for below poverty-level wages; employers can impose long, short, or unpredictable schedules that interfere with employees' family responsibilities, and they are not obligated to provide information sufficient to enable employees to plan lives and careers. Redefining what constitutes the minimum standards of fair treatment is long overdue.

In early 2007, the federal minimum wage remained at a scant $5.15 per hour, unchanged since 1996, and held a real value that was lower than it was in 1951. Although teenagers are commonly considered to be the typical minimum wage workers, the reality is that nearly one in two (47%) minimum wage workers is older than age 25, and two in five (39%) work full time (Haugen 2003; Shulman 2005). As we write this book, a newly elected Democratic Congress passed a bill that will raise the minimum wage to $7.25 per hour by 2009. This standard approximates that advocated by 650 leading economists (including five Nobel Laureates) (Economic Policy Institute 2006). However, in contrast to the recommendations offered by these leading economists, the bill does not tether the increased minimum standard to inflation adjustments, so the minimum wage will continue to be subject to the vagaries of political will.

During the Great Depression, the problem of unemployment was exacerbated because those who did have jobs were expected to work long hours, which in turn deprived others of the opportunity to work. The Fair Labor Standards Act responded by implementing overtime provisions that penalized employers for working their employees beyond 40 hours per week. By mandating time-and-a-half compensation for overtime, the act stimulated a reduction in work hours, as well as dispersed work opportunity. But because full-time employees are typically the only ones who receive benefits, and because the costs of benefits like health care have increased substantially, the penalty for overworking employees has decreased in the new economy. And because of changing opportunity structures, many more employees in the new economy are exempt from overtime provisions, as they labor in salaried positions and within organizational cultures that treat long hours as desirable and normal.

Today, the 40-hour work week plus time-and-a-half pay for overtime is a taken-for-granted arrangement, but one that needs to change. Although the current threshold may have been a reasonable expectation for a husband/breadwinner–wife/homemaker arrangement, today many dual-earner couples work a combined 80 to 100 hours per week (and sometimes more), leaving them frazzled and exhausted. One approach to fixing this problem can be found in France, a country that reduced the threshold for overtime eligibility to 35 hours per week. The result of this legislation has been to increase work opportunity and to decrease the level of stress shouldered by working families. One study of the impact of the French law found that nearly two in three workers responded that it had made it easier for them to combine work life with their family life (Fagnani and Letablier 2004). In 2007, the newly elected Nicolas Sarkozy, who ran for president on a *conservative* platform, stated no plans to fundamentally rework the 35-hour labor standard. Alternately, rather than lowering the threshold at which overtime begins, the penalty for overworking employees could be increased—perhaps from time-and-a-half to double-time compensation.

A Fair Labor Standards Act for the New Economy needs to address the issue of overwork in other ways as well. For example, to bring the country more in line with the treatment of workers in Europe, Americans would need the opportunity to take 4 to 5 weeks of paid vacation a year. Current labor standards need to go far beyond the 12 weeks of unpaid leave provided by the Family and Medical Leave Act and at a minimum, the current standard needs to be modified to accommodate paid leaves of absence (which in turn will make family leave available to all workers). The variety of family leave and family supports enacted in Europe in the past 30 years provide valuable lessons to inform efforts to reshape American family leave policy (see Gornick and Meyers 2003; Hobson 2002; Kelly 2006; Pfau-Effinger 2004).

Finally, a Fair Labor Standards Act for the New Economy needs to address the issue of job insecurities. In part, adjustments of family leave and overtime provisions may help to reduce insecurity by discouraging underemployment and forced "voluntary" exits from the labor force. Again, lessons from Europe may help to craft this legislation, as well as anticipate its impact on worker performance. In France, for example, complex "licenciement" laws regulate the conditions under which employees can be fired. To terminate a worker, companies commonly have to demonstrate that the employee could not be retained, show that the company cannot afford to keep the job in existence, or prove that the employee is incompetent. As a result, when employees underperform, rather than firing them outright (as is commonly done in America), French employers sometimes pay them to disappear graciously, and even go so far as to keep problematic employees on the books, but not assign them important work (Smith 2006). Although this type of legislation could have the result of keeping greater numbers of "deadbeat" employees on the payroll, French workers are actually more productive (as measured by productivity per hour worked) than American employees are (Krugman 2005). Whether American culture could embrace such commitments remains to be seen, but the key to winning the debate will be concerted effort to shift the discussion from employer rights to worker rights (Shaiken 1984), and from "you're on your own" to "we're in this together" (Bernstein 2006).

Expanding Equities in Affluence and Opportunity

Some would argue that the U.S. government should not be in the business of equalizing opportunity, but the fact is that it has been active in this capacity for some time. Were it not for government intervention, rural America would not have paved roads or be electrified, those living downstream or downwind would drink and breathe far stronger toxins, and the public airwaves would not exist. The passage of the Civil Rights Act, Equal Pay Act, and Americans With Disabilities Act, as well as landmark court decisions such as *Brown v. Board of Education,* all demonstrate commitments to equalize opportunities. And government is the source of a number of entitlements— resources that all individuals have access to, regardless of their attachment to the labor force. One such entitlement is public education, which is available to all citizens through the 12th grade. Social Security, although often presented as an "insurance" to which individuals "contribute," is also an entitlement, in that it provides incomes to older citizens and the disabled through the tax revenues generated by the current generation of workers.

At the domestic level, the primary impediment to the provision of collective supports is not scarcity of available resources but rather, the *distribution* of those resources. In 2004, the most affluent fifth of the population received *half* of the nation's collective income. In contrast, the poorest fifth of the population received only a 4% share (*Statistical Abstracts of the United States* 2006). Most other developed nations have implemented systems to offer quality health care, child care, and educational opportunities. Why can't Americans, who live in one of the world's most prosperous economies, have the same resources?

Health Care. In 2003, 45 million Americans (one in seven) lacked health insurance, and those who did have health insurance were most likely to receive it through their jobs (DeNavas-Walt, Proctor, and Mills 2005). Workers in secondary labor market jobs are seldom provided health insurance coverage, leaving a major segment of the workforce unprotected. And for those in the primary labor market, health security depends on maintaining a stable attachment to jobs that are increasingly unstable. Although the Consolidated Omnibus Budget Reconciliation Act of 1985 (COBRA) gave displaced workers the right to continue to purchase insurance through their previous health care plans, many unemployed workers are unable to afford insurance, as budgets stretch to the breaking point (Warren and Tyagi

2003). As fewer workers are covered by employer-sponsored health plans, as health care costs continue to escalate, and as these costs contribute to the export of jobs, reworking the health care system to provide affordable insurance to all Americans needs to be a national priority.

Child Care. Access to quality child care is uneven, expensive, and often inadequate (Children's Defense Fund 2005; Heymann 2000). Today, low-income workers commonly lack access to good child care arrangements, and the middle class feels the financial pinch as well. One possibility is to follow the model offered by Finland and France, which have publicly funded high-quality day care facilities to help working parents remain in their jobs (Pfau-Effinger 2004). Instituting such changes in America will require demonstrating that child care centers can offer care that is as good (or better) than that which is received in the home. Establishing these centers will require directing public resources to constructing buildings, hiring trained personnel, and compensating them at the professional-level wages that care work deserves.

Ironically, as the need to care for disadvantaged children has increased, welfare reform efforts in the United States have undermined the prospects that poor single mothers can nurture their children (Crouter and Booth 2004). If the poor are expected to work, then jobs should pay wages sufficient to support families, and workers should have access to the resources to ensure that children have reasonable care and supervision. Lacking those resources, or the opportunity or ability to work, they should have sufficient public support to enable a reasonable standard of living.

Education. International comparisons of educational attainment are commonly used to make an argument that American public schools are not equipping students with the skills for work in the global economy. The reality is more complex. Those students who are fortunate enough to attend the better schools score as well—and usually higher—than students in nearly all other countries. However, students in disadvantaged school districts perform at low levels. Remedying class and racial divides will involve increasing resources to underfunded schools and combating racial and economic segregation. Beyond this obvious concern, there needs to be a national discussion about how to make a quality college education accessible to all members of society—especially to those students who lack financial resources (Kozol 2006; Marshall and Tucker 1992).

The Role of International Organizations

The latter part of the 20th century witnessed the creation of international trade agreements covering multiple countries. The most famous example is the North American Free Trade Agreement (NAFTA), but there are others around the world, including the creation of the European Union (EU), which goes well beyond a simple trade agreement. These arrangements are fueled by the belief that free trade promotes economic growth and that allowing unlimited capital mobility and tariff-free exchanges will boost all the economies involved, create employment, and generate profits. Because of this motivation, most attempts to assess the impact of these arrangements have focused on job creation. Critics argue that agreements like NAFTA encourage employers to shift employment to low-wage areas, but advocates respond that low-end jobs leave the United States, and new managerial and professional jobs are created as the result of business growth (Bair and Gereffi 2003; Gereffi 1994). Sorting out precisely what the economic impact of NAFTA has been has proven to be tremendously difficult.

NAFTA was accompanied by multinational side agreements on labor and environmental issues, which were intended to provide protections for workers and for the environment in all of the countries signing the agreement. Advocates for organized labor and for environmental causes complain that these have proven largely ineffective (Tseming 2004). Indeed, there is considerable fear that free trade will lead to a global race to the bottom, with countries reducing standards to stay competitive. But it could have been, and could still be, otherwise. If the best labor practices from each country became the standard in all of the participating nations, and if the enforcement mechanisms built into agreements like NAFTA were used effectively, this would improve the working conditions of workers throughout the region. There is much discussion of this within the EU, and in some ways, this may be happening—for example, in the efforts to standardize the maximum number of hours of work. The EU has also used its influence to require improved labor standards in nonmember countries with which it makes bilateral agreements. In other words, international economic agreements *can* become mechanisms for raising labor standards on a global scale.

Conversely, an opposite strategy of withdrawing trade can be used to bring countries with poor labor and environmental records into line. Should the United States "interfere" with the domestic policies of other countries in such a manner? The answer is that it has openly (as well as covertly) done so for some time. Currently, the U.S. government prohibits American businesses from trading with Myanmar (Burma), Ivory Coast, Sudan, and Cuba over human rights issues. And although the exploitation of workers abroad appears external to U.S. interests, it is one of the chief factors that stimulates the migration of jobs from the United States to developing countries. Here is a nexus where activist groups and unions can play a crucial role in influencing discussion of the conduct of international trade relations.

Human rights abuses at work come in a variety of forms, but one of the most devastating is child labor. UNICEF estimates that 246 million children work in the global economy and that approximately half are employed in hazardous environments. In Benares, India, as many as 200,000 children work in virtual bondage for manufacturers producing saris and carpets (Davis 2006). In Bangladesh, two-thirds of children aged 12 to 16 work and labor on average 8 to 10 hours per day (Delap 2001). Throughout the developing world, children are employed in a wide variety of endeavors in the apparel, shoe, fishing, rug, and construction industries. They are also major contributors to informal economies, working as (among other occupations) prostitutes, servants, and firewood collectors.

Some lessons can be learned from considering why child labor declined in Europe and America but persists in the developing world.[6] In the mid-19th century, children composed as much as 60% of the workforces in the factories that emerged in the wake of the Industrial Revolution (Gratton and Moen 2004). In the United States, children contributed one-quarter to one-third of the typical household's income in the early 20th century, which was more than the average contribution of wives (Cunningham 2000). Looking back at the decline of child labor in America reveals two forces at play. One force was cultural—children were redefined as needing protection (Aries 1965; Zelizer 1994). But equally and perhaps even more important were the changing rewards received from work, which made it possible for families to remove their children from the labor force and still maintain a reasonable quality of life (Cunningham 2000; Edmonds 2005). In developing nations, where per capita earnings are in the hundreds of dollars per year, a child's income can be critical to family survival. If standards of living rise in the developing world, we would anticipate child labor declines similar to those that occurred in the early 20th century in the United States. Reaching this goal will almost certainly require developing more effective means of regulating of global supply chains.

One would hope that international agencies, such as the United Nations, would be well positioned to regulate global work, but this does not appear to be the case. Consider, for example, the United Nations' response to the problem of child labor in the new economy. In 1989, the United Nations ratified the following resolution:

Children have the right to be protected from economic exploitation and from performing any work that is likely to be hazardous or to interfere with the child's education, or to be harmful to the child's health or physical, mental, spiritual, moral or social development. Additionally, states must (a) Provide for a minimum age or minimum ages for admission to employment; (b) Provide for appropriate regulation of the hours and conditions of employment; (c) Provide for appropriate penalties or other sanctions to ensure the effective enforcement of the present article. (United Nations 1989)

Although it condemns child labor, enforcement of this resolution has been hamstrung by vague standards on who is (or is not) a child, as well as how many hours and what conditions are "appropriate." Even in countries like India that have laws prohibiting the use of child labor, lax oversight of employers is common. As one analyst noted, it is ironic that India forbids child labor in the carpet weaving industry, yet the Indian government itself runs a training program for carpet weavers and recruits children as young as age 6. And although the world community condemns child labor, global investment practices (discussed later) continue to provide millions of dollars of funds to support the expansion of the carpet industry, with no barriers to prevent the use of child labor (Arat 2002).

The ability and the will of international organizations to go beyond merely symbolic gestures on the problem of labor abuse are questionable. The International Labor Organization (ILO) is the United Nations' branch concerned with labor questions. In principle, the ILO could become a force for improving the conditions of labor on a global scale and for using the United Nation's enforcement authority to combat labor abuses in particular parts of the world. However, the ILO has largely defined itself as a clearinghouse for information, a center for research, and a forum for discussion of the problems of the workplace, but not as an enforcement agency.

Today, there appears to be a greater emphasis on opening trade than on regulating and controlling the terms under which it occurs. Consider, for example, the ways in which the World Bank and International Monetary Fund (IMF) approach the issue of fostering the development of the global economy. Although these are separate organizations formed for different reasons, by the end of the 20th century they shared three common concerns:

1. Crisis management—to provide economic resources to help countries avert financial crises.

2. Transition—to foster countries' efforts to restructure to market-oriented economies.

3. Development—to help the poorest countries obtain resources to integrate themselves into global markets.

Their approach to resolving these problems has been to provide loans to countries in need, but under conditions that require political reforms in support of market-based economies. By providing loans to countries in need, the World Bank and IMF invest in high-risk economies, putting money where it is lacking. But the Structural Adjustment Programs that accompany these loans typically require reduced government supports for workers, privatization, diminished regulation, and a series of other measures that at best require no improvement in the conditions of labor and at worst lead to deterioration. These programs also provide little protection against loans being wasted through political waste and outright corruption. Although one of the IMF's and World Bank's goals is to foster economic development, there is no clear evidence that they have been successful, and some evidence that their lending practices may have even undermined economic development. Such is the case of lending to sub-Saharan Africa. From 1971 to 1995, debts to the IMF and World Bank escalated from $10 billion to $230 billion. As a result, by 1995, the total debts owed by sub-Saharan African countries constituted nearly 70% of their gross national products—leading many societies to introduce austerity budgets and to cut back on needed investments in public education. Lacking the resources to pay these loans, these already poor countries are trapped in debt-borrowing spirals that now impoverish them, rather than help them develop (Babb 2005; Woods 2006).

The United States is in no position to tell other societies that they cannot use child labor, or dictate wages and environmental practices. But it can dictate that, unless specific expectations are met, U.S. investors will be prohibited from engaging in trade with those societies. The first step is to initiate a national discussion of what these standards should be, and to foster a public awareness that it is in the interests of U.S. workers to have such standards enforced. Again, rather than focusing on influencing employers through a "name and shame" strategy, greater prospects for change may emerge from activist groups' efforts to pressure the U.S. government to support workers' movements at home and abroad (Seidman 2007).

The United States could also play a greater role in stimulating positive development. Despite being the major recipient of wealth generated in the global economy, the U.S. government lags far behind most countries in the share of its wealth it redistributes to developing countries. Of equal concern is the expectation that agencies such as the IMF and World Bank increasingly perform the needed development work. In 2004, the United States allotted less than 1% of its budget to foreign aid (only 18 cents per $100 of the country's GDP). The largest share of aid given in 2004 was directed to military and political efforts in the Middle East (Iraq, Israel, and Egypt), and far smaller proportions were directed to true development efforts in impoverished regions of the world such as Africa and Southeast Asia. It also bears noting that the amount of aid being given, as a percentage of the United States' GDP, has declined steadily from the years following World War II to the present (Tarnoff and Nowels 2004).

Conclusion

In this book, we argued that the new economy has many new characteristics but that it has also integrated old practices into the design of jobs and the allocation of resources. Some of the problems evident today are new, but many reflect unresolved problems that emerged in the old economy, failed approaches to those problems, and lagging responses to emerging concerns. Today, America faces challenges analogous to those it confronted at the turn of the 20th century. Old means of working are being discarded, as new technologies and organizational practices are introduced. Previously dominant sectors of the economy are in decline, as new jobs and industries are emerging. And the workforce is changing, as are communities of workers. Along with the new challenges these changes present, there remain persistent concerns—of inequalities, overwork, unemployment, insecurity, and strained schedules.

If left unchanged, the contours of work in the new economy will be all too familiar. The future of work will be characterized by divided opportunities, with workers segregated on the basis of class, gender, race, and nationality. Some workers, particularly women and members of minority groups, will have careers dislodged or never make it onto career tracks at all. Large portions of the workforce will labor in alienating, low-skilled, low-wage jobs. Some will have careers disrupted when they have children; others will keep their careers intact by foregoing having children altogether. Large numbers of workers will experience stress from having too much work, whereas many others will be work-poor and labor in jobs with unpredictable schedules, if they can find work at all. And most workers will labor under conditions of uncertainty and have inadequate resources to plan careers and to weather job loss.

Economies should work for people, not the other way around. The shape of the new economy is not immutable, nor is its future inevitable. One of the lessons learned by looking at work in historical perspective is that people have been able to effect positive change—even when faced with formidable resistance. Throughout this book, we presented ideas for change and illustrations of how these changes have been introduced in other countries. A wide variety of existing examples show how societies and organizations have reduced overwork, fostered employment, expanded flexibilities, increased securities, and equalized opportunities.

Essential to solving all of the issues presented in this book is recognizing that the problem does not rest in a scarcity of wealth or resources. It rests in the ways resources are allocated in a tremendously productive economy.

There is no reason to accept the premise that good jobs and a good life will inevitably be beyond the reach of the hardworking men and women who make it all possible.

Notes

1. Such ordinances commit communities to do business only with employers who pay wages above minimum wage levels and that enable a family to live at a level above the federally defined poverty line.

2. Authors' analysis of the General Social Survey.

3. For example, Milkman notes that the AFL tries to get all (or most) of the employers in an area to agree to labor standards so that a single employer's threat to move or close down is less effective.

4. In 1995, when John Sweeney became leader of the AFL-CIO, he promised to rebuild the union movement by prioritizing organizing activities. The new regime did have some successes, most notably the Justice for Janitors campaign. However, the AFL-CIO's efforts in this regard may have stalled. Careful analysis of contemporary unions shows that only a few of the AFL-CIO's affiliates have made organizing new groups of workers a priority (Bronfenbrenner and Hickley 2003). In 2005, the AFL-CIO suffered a major split. A number of unions, including some of its largest and most vigorous members (SEIU, AFSCME) left to form their own federation (Change to Win) dedicated to more aggressive organizing. It remains to be seen whether this will divide and weaken the labor movement, or increase new organizing activity (Moberg 2005).

5. Ben & Jerry's was, however, recently purchased by an international conglomerate. There has been an effort to preserve the company's socially conscious image, but it remains to be seen whether this can be sustained over the long haul.

6. One notable contradiction in the United States is the continued acceptance of child labor on farms, where they are engaged in work that is dangerous, backbreaking, and largely unenjoyable. Because farm workers are largely excluded from the protections of the Fair Labor Standards Act, children at age 12 (and in some instances even younger) can legally be employed for wages lower than the conventional minimum wage and with no limit on the number of hours the child may be compelled to work each day. Limited regulation of farm work and lax enforcement of the laws that do exist result in exposure to dangerous equipment and chemical contamination (Tucker 2000).

Appendix

Legislative and Regulatory
Timeline of Worker Rights and
Protections in the United States

1884

Federal Labor Bureau, the predecessor of the Bureau of Labor Statistics, was established by the Hopkins Act.

1891

Kansas established the first state prevailing wage law.

1903

Department of Commerce and Labor was established by an act of Congress.

1912

Massachusetts adopted first minimum wage law for women and minors.

1913

U.S. Department of Labor was established by an act of Congress. It included the Bureau of Labor Statistics, the Bureau of Immigration and Naturalization, and the Children's Bureau.

Adapted from Bureau of Labor Statistics, http://www.bls.gov/opub/rtaw/pdf/lrtime.pdf

1914

Clayton Act limited the use of injunctions in labor disputes and provided that picketing and other union activities should not be considered unlawful.

1916

First federal child labor law was signed, but was later struck down by the U.S. Supreme Court.

1920

Women's Bureau was established as a federal agency to represent the needs of wage-earning women in the public policy process.

1926

Railway Labor Act required railroad employers to not discriminate against employees for joining a union and provided for the rights of workers to collectively bargain.

1931

Davis-Bacon Act provided for the payment of prevailing wage rates to laborers and mechanics employed by contractors and subcontractors on public construction.

1932

Norris-LaGuardia Act outlawed "yellow dog" contracts that required workers to sign away their rights to join unions as a precondition for employment.

1933

Wagner-Peyser Act created U.S. Employment Service in Department of Labor.

1935

Federal Social Security Act provided a nationwide system of social insurance to protect wage earners and their families in old age, in the event of illness and disability, and in the event of economic hardship. Many key social programs of the 20th century developed from this act including Social Security, Unemployment Insurance, and Aid to Families With Dependent Children (AFDC). National Labor Relation (Wagner) Act established the first national policy of protecting the rights of workers to organize and elect their representatives for collective bargaining purposes.

1936

Public Contracts (Walsh-Healy) Act set labor standards on government contracts requiring the manufacture or purchase of materials.

1938

Fair Labor Standards Act set minimum wage, maximum hours, and time pay, as well as equal pay and child labor standards.

1947

Labor-Management Relations (Taft-Hartley) Act reiterated policies protecting rights of workers to organize and elect union representatives. But this act also placed checks on union tactics, including secondary strikes and organizing procedures, and as such is commonly viewed by labor activists as a step backward in the labor movement.

1949

An amendment to the Fair Labor Standards Act directly prohibited child labor for the first time.

Courts decide that benefits are subject to collective bargaining.

1958

Welfare and Pension Disclosure Act required administration of health insurance, pension, and supplementary unemployment compensation plans to file plan descriptions and annual financial reports with the Secretary of Labor.

1959

Labor-Management Reporting (Landrum-Griffin) Act prohibited improper activities by labor and management, such as secondary boycotts; provided certain protection for the rights of union members; and required filing of certain financial reports by unions and employers.

1962

Manpower Development and Training Act required federal government to determine manpower requirements and resources and to "deal with the problems of unemployment resulting from automation and technological changes and other types of unemployment."

1963

Equal Pay Act prohibited wage differentials based on sex for workers covered by the Fair Labor Standards Act.

1964

Title VII of the Civil Rights Act established U.S. Equal Employment Opportunity Commission to enforce federal statutes prohibiting employment discrimination.

1965

Medicare established under Social Security.

McNamara-O'Hara Service Contract Act provided wage standards for employees performing work on federal service contracts.

1967

Age Discrimination in Employment Act made it illegal to discharge, refuse to hire, or otherwise discriminate against persons ages 40 to 65.

1969

Federal Coal Mine Health and Safety Act protected the health and safety of the nation's coal miners.

1970

Occupational Safety and Health Act (OSHA) placed certain duties on employers and employees to assure safe and healthful working conditions.

1974

Employer Retirement Income Security Act (ERISA) imposed standards on employer-provided benefit plans. Act was designed to protect the security of pension promises made by private sector firms.

1978

Pregnancy Discrimination Act required employee benefit programs to treat pregnancy in the same way as illnesses.

1982

Job Training Partnership Act (JPTA) prepared youths and adults facing serious barriers to employment by providing job training and other services that would result in increased earnings, increased education and occupational skills, and decreased welfare dependency.

1985

Consolidated Omnibus Budget Reconciliation Act (COBRA) required employers that provide health care benefits to continue such benefits to formerly covered individuals for a period of time after employer coverage ends.

1989

Worker Adjustment and Retraining Notification Act (WARN) provided protection to workers, their families, and their communities, by requiring employers to provide notification 60 calendar days in advance of plant closings and mass layoffs.

1990

Americans with Disabilities Act (ADA) established a clear and comprehensive prohibition of discrimination on the basis of disability.

1993

Family and Medical Leave Act (FMLA) mandated employers to provide many types of workers up to 12 weeks of unpaid time off for worker and family medical purposes.

1996

The Personal Responsibility and Work Opportunity Reconciliation Act introduced work expectations for the receipt of welfare, replacing the earlier program of Aid to Families With Dependent Children (AFDC) with Temporary Assistance for Needy Families (TANF). Especially important changes were a maximum 5 years' cumulative support and the introduction of expectations that welfare recipients work (or prepare to work).

1998

The Workforce Investment Act (WIA) focused on the needs of companies and how to make industries more productive by providing services that increase the number of jobs and the placement of individuals into those jobs.

1999

Ticket to Work and Work Incentives Act (TWWIA) provided health care and employment preparation and placement services to individuals with disabilities.

References

Abelson, Reed. 1999. "A Push From the Top Shatters a Glass Ceiling." *New York Times* (August 20), p. 1.

Alba, Richard, and Victor Nee. 2003. *Remaking the American Mainstream: Assimilation and Contemporary Immigration*. Cambridge, MA: Harvard University Press.

Altonji, Joseph, and Jennifer Oldham. 2003. "Vacation Laws and Annual Work Hours." *Federal Reserve Bank of Chicago Economic Perspectives*; 19–29.

Altucher, Kristine, and Lindy B. Williams. 2003. "Family Clocks: Timing Parenthood." Pp. 49–59 in *It's About Time: Career Strains, Strategies, and Successes*, edited by Phyllis Moen. Ithaca, NY: Cornell University Press.

Aman, Carolyn, and Paula England. 1997. "Comparable Worth: When Do Two Jobs Deserve the Same Pay?" In *Subtle Sexism: Current Practice and Prospects for Change*, edited by Nijole Benokraitis. Thousand Oaks, CA: Sage.

American Academy of Pediatrics. 2005. "Breastfeeding and the Use of Human Milk." *Pediatrics* 115: 496–506.

Amott, Teresa, and Julie Matthaei. 1996. *Race, Gender, and Work: A Multicultural Economic History of Women in the United States*. Boston: South End Press.

Anderson, Elijah. 1999. *Code of the Street: Decency, Violence, and the Moral Life of the Inner City*. New York: Norton.

Appelbaum, Eileen, and Rosemary Batt. 1992. *The New American Workplace: Transforming Work Systems in the United States*. Ithaca, NY: ILR Press.

Appelbaum, Steven, Robert Simpson, and Barbara Shapiro. 1987. "The Tough Test of Downsizing." *Organizational Dynamics* 16: 68–79.

Arat, Zehra F. 2002. "Analyzing Child Labor as a Human Rights Issue: Its Causes, Aggravating Policies, and Alternative Proposals." *Human Rights Quarterly* 24: 177.

Arenson, Karen. 2005. "N.Y.U. Moves to Disband Graduate Students Union." *New York Times* (June 17), p. 2.

Aries, Phillipe. 1965. *Centuries of Childhood: A Social History of Family Life*. New York: Vintage.

Armbruster Sandoval, Ralph. 2005. "Workers of the World Unite? The Contemporary Anti-Sweatshop Movement and the Struggle for Social Justice in the Americas." *Work and Occupations* 32: 464–485.

Armstrong-Stassen, Marjorie. 1993. "Survivors' Reactions to a Workforce Reduction." *Canadian Journal of Administrative Sciences* 10: 334–344.

———. 1998. "The Effect of Gender and Organizational Level on How Survivors Appraise and Cope with Organizational Downsizing." *The Journal of Applied Behavioral Science* 34: 125–142.

Arthur, Michael B., and Denise M. Rousseau. 1996. *The Boundaryless Career: A New Employment Principle for a New Organizational Era.* New York: Oxford University Press.

Astin, Helen S., and Jeffrey F. Milem. 1997. "The Status of Academic Couples in U.S. Institutions." Pp. 128–155 in *Academic Couples: Problems and Promises,* edited by Marianna A. Ferber and Jeffrey W. Loeb. Chicago: University of Illinois Press.

Babb, Sarah. 2005. "The Social Consequences of Structural Adjustment: Recent Evidence and Current Debates." *Annual Review of Sociology* 31: 199–222.

Bailyn, Lotte, Ann Bookman, Mona Harrington, and Thomas Kochan. 2005. "Work-Family Interventions and Experiments: Workplaces, Communities, and Society." In *The Work and Family Handbook: Multi-Disciplinary Perspectives, Methods, and Approaches,* edited by Marcie Pitt-Catsouphes, Ellen Ernst Kossek, and Stephen Sweet. Mahwah, NJ: Erlbaum.

Bair, Jennifer, and Gary Gereffi. 2003. "Upgrading, Uneven Development, and Jobs in the North American Apparel Industry." *Global Networks* 3: 143–169.

Bandura, Albert. 1982a. "Self-Efficacy Mechanism in Human Agency." *American Psychologist* 37: 122–147.

———. 1982b. "The Self and Mechanisms of Agency." Pp. 3–39 in *Psychological Perspectives on the Self,* edited by J. Suls. Hillsdale, NJ: Erlbaum.

Barnett, Rosalind Chait, and Karen C. Gareis. 2000. "Reduced-Hours Employment: The Relationship Between Difficulty of Trade-Offs and Quality of Life." *Work and Occupations* 27: 168–187.

Bartlett, Donald, and James Steele. 1992. *America: What Went Wrong?* Kansas City: Andrews and McMeel.

Batt, Rosemary, Virginia Doellgast, Hyunji Kwon, Mudit Nopany, Priti Nopany, and Anil da Costa. 2005. *The Indian Call Centre Industry: National Benchmarking Report Strategy, HR Practices, and Performance* (CAHRS Working Paper #05-07). Ithaca, NY: Cornell University, School of Industrial and Labor relations, Center for Advanced Human Resources Studies. Retrieved from http://digitalcommons.ilr.cornell.edu/cahrswp/7

Baumol, William J., Alan S. Blinder, and Edward N. Wolff. 2003. *Downsizing in America: Reality, Causes, and Consequences.* New York: Russell Sage Foundation.

Beck, Ulrich. 2000. *The Brave New World of Work.* Cambridge, UK: Polity Press.

Becker, Gary S. 1981. "Division of Labor in Households and Families." Pp. 30–53 in *A Treatise on the Family,* edited by Gary S. Becker. Cambridge, MA: Harvard University Press.

Becker, Penny Edgell, and Phyllis Moen. 1999. "Scaling Back: Dual-Earner Couples' Work-Family Strategies." *Journal of Marriage and the Family* 61: 995–1007.

Bell, Daniel. 1973. *The Coming of Post-Industrial Society.* New York: Basic Books.

Bellah, Robert, Richard Madsen, William Sullivan, Ann Swidler, and Steven Tipton. 1985. *Habits of the Heart: Individualism and Commitment in American Life*. New York: Harper & Row.

Bem, Sandra. 1993. *The Lenses of Gender: Transforming the Debate on Gender Inequality*. New Haven, CT: Yale University Press.

Benokraitis, Nijole. 1997. "Sex Discrimination in the 21st Century." In *Subtle Sexism: Current Practice and Prospects for Change*, edited by Nijole Benokraitis. Thousand Oaks, CA: Sage.

Bensman, David, and Roberta Lynch. 1987. *Rusted Dreams: Hard Times in a Steel Community*. Berkeley: University of California Press.

Bergman, Helena, and Barbara Hobson. 2002. "Compulsory Fatherhood: The Coding of Fatherhood in the Swedish Welfare State." Pp. 92–124 in *Making Men Into Fathers: Men, Masculinities and the Social Politics of Fatherhood*, edited by Barbara Hobson. Cambridge, UK: Cambridge University Press.

Bernhardt, Annette, Martina Morris, Mark Handcock, and Marc Scott. 2001. *Divergent Paths: Economic Mobility in the New American Labor Market*. New York: Russell Sage Foundation.

Bernstein, Jared. 2006. *All Together Now: Common Sense for a Fair Economy*. San Francisco: Berrett-Koehler.

Bertrand, Marianne, and Sendhil Mullainathan. 2004. "Are Emily and Greg More Employable Than Lakisha and Jamal? A Field Experiment on Labor Market Discrimination." *The American Economic Review* 94: 991–1013.

Bielby, D. D. 1992. "Commitment to Work and Family." *Annual Review of Sociology*: 281–302.

Bielby, William T., and Denise D. Bielby. 1992. "I Will Follow Him: Family Ties, Gender-Role Beliefs, and Reluctance to Relocate for a Better Job." *American Journal of Sociology* 97: 1241–1267.

Blair-Loy, Mary. 2003. *Competing Devotions: Career and Family Among Women Executives*. Cambridge, MA: Harvard University Press.

Blau, Francine, and Lawrence Kahn. 2006. "The U.S. Gender Pay Gap in the 1990s: Slowing Convergence." In *Industrial Relations Section Working Paper*, vol. 508. Princeton, NJ: Princeton University.

Blowfield, Michael. 2005. "Corporate Social Responsibility: Reinventing the Meaning of Development?" *International Affairs* 81: 515–524.

Bluestone, Barry, and Bennett Harrison. 1982. *The Deindustrialization of America: Plant Closings, Community Abandonment, and the Dismantling of Basic Industry*. New York: Basic Books.

Bonacich, Edna. 1972. "A Theory of Ethnic Antagonism: The Split Labor Market." *American Sociological Review* 37: 547–59.

Bookman, Ann. 2004. *Starting in Our Own Backyards: How Working Families Can Build Community and Survive the New Economy*. New York: Routledge.

Boris, Eileen, and Carolyn Lewis. 2005. "Caregiving and Wage-Earning: A Historical Perspective on Work and Family." Pp. 73–98 in *The Work and Family Handbook: Multi-Disciplinary Perspectives, Methods and Approaches*, edited by Marcie Pitt-Catsouphes, Ellen Ernst Kossek, and Stephen Sweet. Boston: Erlbaum.

Borjas, George, Richard Freeman, Lawrence Katz, John DiNardo, and John Abowd. 1997. "How Much Do Immigration and Free Trade Affect Labor Market Outcomes?" *Brookings Papers on Economic Activity* 1997: 1–90.

Bourdieu, Pierre. 1984. *Distinction: A Social Critique of the Judgment of Taste.* Cambridge, MA: Harvard University Press.

———. 1986. "The Forms of Capital." In *Handbook of Theory and Research for the Sociology of Education,* edited by John Richardson. New York: Greenwood Press.

Bourgois, Phillippe. 1995. *In Search of Respect: Selling Crack in El Barrio.* New York: Cambridge University Press.

Bowen, William, and Derek Bok. 1998. *The Shape of a River: Long-Term Consequences of Considering Race in College and University Admissions.* Princeton, NJ: Princeton University Press.

Boydston, Jeanne. 1990. *Home and Work: Housework, Wages, and the Ideology of Labor in the Early Republic.* New York: Oxford University Press.

Bradsher, Keith. 2002. "A High-Tech Fix for One Corner of India." *New York Times* (December 27), C1.

Braverman, Harry. 1974. *Labor and Monopoly Capital.* New York: Monthly Review Press.

Breen, T. H. 1985. *Tobacco Culture: The Mentality of the Great Tidewater Planters on the Eve of the Revolution.* Princeton, NJ: Princeton University Press.

Brockner, Joel. 1990. "Scope of Justice in the Workplace: How Survivors React to Co-Worker Layoffs." *Journal of Social Issues* 46: 95–106.

Brockner, Joel, Batia M. Wiesenfeld, and Christopher L. Martin. 1995. "Decision Frame, Procedural Justice, and Survivors' Reactions to Job Layoffs." *Organizational Behavior and Human Decision Processes* 63: 59–68.

Brockner, Joel, Steven Grover, Thomas Reed, Rocki Lee DeWitt, and Michael O'Malley. 1987. "Survivors' Reactions to Layoffs: We Get by With a Little Help From Our Friends." *Administrative Science Quarterly* 32: 526–541.

Broder, John. 2006. "Immigrants and the Economics of Hard Work." *New York Times* (April 2), p. WK3.

Brody, Hugh. 2002. *Maps and Dreams.* New York: Pantheon.

Broman, Clifford L., V. Lee Hamilton, and William S. Hoffman. 1990. "Unemployment and Its Effects on Families: Evidence From a Plant Closing Study." *American Journal of Community Psychology* 18: 643–659.

Bronfenbrenner, Kate, and Robert Hickley. 2004. "Changing to Organize: A National Assessment of Union Strategies." Pp. 17–61 in *Rebuilding Labor: Organizers and Organizing in the New Labor Movement,* edited by Ruth Milkman and Kim Voss. Ithaca, NY: ILR Press.

Brooks, David. 2006. "The Education Gap." *New York Times* (September 25), p. 11.

Brooks, Michael. 2001. "Law Enforcement Physical Fitness Standards and Title VII." *FBI Law Enforcement Bulletin* 70: 26–32.

Browne, Irene, and Ivy Kennelly. 1999. "Stereotypes and Realities: Images of Black Women in the Labor Market." Pp. 302–326 in *Latinas and African American Women at Work: Race, Gender, and Economic Inequality,* edited by Irene Brown. New York: Russell Sage Foundation.

Buchanan, Patrick. 2006. *State of Emergency: The Third World Invasion of America.* New York: Thomas Dunne Books.

Budig, Michelle. 2006. "Intersections on the Road to Self-Employment: Gender, Family and Occupational Class." *Social Forces* 84: 2223–2239.

Burawoy, Michael. 1979. *Manufacturing Consent: Changes in the Labor Process Under Monopoly Capitalism.* Chicago: University of Chicago Press.

Bureau of Labor Statistics. 2004. "Employee Tenure in 2004." Retrieved from http://www.bls.gov/news.release/archives/tenure_09212004.pdf

———. 2006a. "Issues in Labor Statistics: Women Still Underrepresented Among Highest Earners," Summary 06–03, U.S. Department of Labor. Washington, DC: U.S. Department of Labor. Retrieved from http://www.bls.gov/opub/ils/pdf/opbils55.pdf

———. 2006b. *Occupational Outlook Handbook.* Washington, DC: U.S. Department of Labor.

———. 2007a. "Union Members Summary," vol. 07–0113. U.S. Department of Labor. Washington, DC: U.S. Department of Labor. Retrieved from http://www.bls.gov/news.release/union2.nr0.htm

———. 2007b. "American Time Use Survey—2006 Results," vol. 07-0930. Retrieved from http://www.bls.gov/news.release/pdf/atus.pdf

Burke, Ronald J., and Esther Greenglass. 1999. "Work-Family Conflict, Spouse Support and Nursing Staff Well-Being During Organizational Restructuring." *Journal of Occupational Health Psychology* 4(4): 327–336.

Buss, Terry F., and F. Stevens Redburn. 1987. "Plant Closings: Impacts and Responses." *Economic Development Quarterly* 1: 170–177.

Butler, John Sibley, and Cedric Herring. 1991. "Ethnicity and Entrepreneurship in America: Toward an Explanation of Racial and Ethnic Group Variations in Self-Employment." *Sociological Perspectives* 34: 79–94.

Callahan, Raymond. 1962. *Education and the Cult of Efficiency.* Chicago: University of Chicago Press.

Capelli, Peter. 1997. *The New Deal at Work.* Cambridge, MA: Harvard Business School Press.

———. 1999. "Career Jobs Are Dead." *California Management Review* 42: 168–179.

Card, David. 1990. "The Impact of the Mariel Boatlift on the Miami Labor Market." *Industrial and Labor Relations Review* 43: 245–247.

———. 2001. "Immigrant Inflows, Native Outflows, and the Local Labor Market Impacts of Higher Immigration." *Labor Economics* 19: 22–64.

Carr, Deborah. 1996. "Two Paths to Self-Employment? Women's and Men's Self-Employment in the United States, 1980." *Work and Occupations* 23: 26–53.

Carter, Craig R. 2004. "Purchasing and Social Responsibility: A Replication and Extension." *Journal of Supply Chain Management: A Global Review of Purchasing & Supply* 40: 4–16.

Castells, Manuel. 2000. *The Rise of the Network Society.* Oxford, UK: Blackwell.

Central Intelligence Agency. 2005. *The World Factbook.* Washington, DC: Central Intelligence Agency.

Chakravartty, Paula. 2001. "Flexible Citizens and the Internet: The Global Politics of Local High-Tech Development in India." *Emergences: Journal for the Study of Media & Composite Cultures* 11: 69–88.

Chandler, Alfred. 1990. *Scale and Scope: The Dynamics of Industrial Capitalism.* Cambridge, MA: Belknap Press.

Chase-Dunn, Christopher, Yukio Kawano, and Benjamin D. Brewer. 2000. "Trade Globalization since 1795: Waves of Integration in the World-System." *American Sociological Review* 65: 77–95.

Children's Defense Fund. 2005. *State of America's Children 2005.* Washington, DC: Children's Defense Fund.

Chodorow, Nancy. 1978. *The Reproduction of Mothering.* Berkeley: University of California Press.

Christensen, Kathleen, and Graham L. Staines. 1990. "Flextime: A Viable Solution to Work/Family Conflict?" *Journal of Family Issues* 11: 455–476.

Chronicle of Higher Education Almanac. 2004. "College Enrollment by Age of Students, Fall 2002." *Chronicle of Higher Education,* August 27, 2004, pp. 16.

Clawson, Dan. 2003. *The Next Upsurge: Labor and the New Social Movements.* Ithaca, NY: ILR Press.

Cohen, Barney. 2003. "Urban Growth in Developing Countries: A Review of Current Trends and a Caution Regarding Existing Forecasts." *World Development* 32: 23–51.

Cohen, Ruth Schwartz. 1985. *More Work for Mother: The Ironies of Household Technology From the Open Hearth to the Microwave.* New York: Basic Books.

Collins, Sharon. 1997. "Black Mobility in White Corporations: Up the Corporate Ladder but Out on a Limb." *Social Problems* 44: 55–67.

Committee on Maximizing the Potential of Women in Academic Science and Engineering, National Academy of Sciences, National Academy of Engineering, and Institute of Medicine. 2006. *Beyond Bias and Barriers: Fulfilling the Potential of Women in Academic Science and Engineering.* Washington DC: National Academies Press.

Coontz, Stephanie. 1992. *The Way We Never Were: American Families and the Nostalgia Trap.* New York: Basic Books.

———. 1997. *The Way We Really Are: Coming to Terms with America's Changing Families.* New York: Basic Books.

Cornelius, Wayne. 2005. "Controlling 'Unwanted' Immigration: Lessons From the United States, 1993–2004." *Journal of Ethnic and Migration Studies* 31: 775–794.

Correll, Shelley. 2001. "Gender and the Career Choice Process: The Role of Biased Self-Assessments." *American Journal of Sociology* 106: 1691–1730.

———. 2004. "Constraints into Preferences: Gender, Status, and Emerging Career Aspirations." *American Sociological Review* 69: 93–113.

Costa, Dora. 2000. "Hours of Work and the Fair Labor Standards Act: A Study of Retail and Wholesale Trade, 1938–1950." *Industrial and Labor Relations Review* 53: 648–664.

Cottle, Thomas J. 2001. *Hardest Times: The Trauma of Long-Term Unemployment.* Westport, CT: Praeger.

Cowie, Jefferson. 2001. *Capital Moves: RCA's 70-Year Quest for Cheap Labor.* New York: New Press.

Crawford, Elizabeth. 2003. "Good to the Last Drop." *Chronicle of Higher Education* 49: A8.

Creamer, Elizabeth, and Associates. 2001. *Working Equal: Academic Couples as Collaborators.* New York: RoutledgeFalmer.

Crittenden, Ann. 2001. *The Price of Motherhood.* New York: Henry Holt.

Crouter, Ann, and Alan Booth. 2004. *Work-Family Challenges for Low-Income Parents and Their Children.* Mahwah, NJ: Erlbaum.

Cunningham, Hugh. 2000. "The Decline of Child Labour: Labour Markets and Family Economies in Europe and North America Since 1830." *Economic History Review* 53: 409–428.

Curry, James. 1993. "The Flexibility Fetish." *Capital and Class* 50: 99–126.

Dark, Taylor. 2001. *The Unions and the Democrats: An Enduring Alliance,* 2nd ed. Ithaca, NY: Cornell University Press.

Darrah, Charles. 2005. "Ethnography and Working Families." Pp. 367–386 in *The Work and Family Handbook: Multi-Disciplinary Perspectives, Methods, and Approaches,* edited by Marcie Pitt-Catsouphes, Ellen Ernst Kossek, and Stephen Sweet. Mahwah, NJ: Erlbaum.

Davis, Mike. 2006. *Planet of Slums.* London: Verso.

Delap, Emily. 2001. "Economic and Cultural Forces in the Child Labour Debate: Evidence from Urban Bangladesh." *Journal of Development Studies* 37: 1.

Dellinger, Kirsten, and Christine Williams. 2002. "The Locker Room and the Dorm Room: Workplace Norms and the Boundaries of Sexual Harassment in Magazine Editing." *Social Problems* 49: 242–257.

DeLong, David. 2004. *Lost Knowledge: Confronting the Threat of an Aging Workforce.* New York: Oxford University Press.

DeNavas-Walt, Carmen, Bernadette Proctor, and Robert J. Mills. 2005. *Income, Poverty, and Health Insurance Coverage in the United States: 2003.* Washington, DC: U.S. Department of Commerce, U.S. Census Bureau.

Department of Homeland Security. 2005. "Yearbook of Immigration Statistics." Department of Homeland Security. Retrieved from http://www.dhs.gov/ximgtn/statistics/publications/yearbook.shtm

Deutsch, Francine. 1999. *Halving It All: How Equally Shared Parenting Works.* Cambridge, MA: Harvard University Press.

Deyo, Frederic. 1989. *Beneath the Miracle: Labor Subordination in the New Asian Industrialism.* Berkeley: University of California Press.

Dill, Bonnie Thorton. 1988. "Our Mothers' Grief: Racial-Ethnic Women and the Maintenance of Families." *Journal of Family History* 13: 415–431.

Dominguez, Silvia, and Celeste Watkins. 2003. "Creating Networks for Survival and Mobility: Social Capital Among African-American and Latin-American Low-Income Mothers." *Social Problems* 50: 111–135.

Dooley, David, Ralph Catalano, and Georjeanna Wilson. 1994. "Depression and Unemployment." *American Journal of Community Psychology* 22: 745–765.

Duncan, Cynthia M. 1992. *Rural Poverty in America*. Westport, CT: Auburn House.

Durkheim, Émile. 1964 [1895]. *The Division of Labor in Society*. New York: Free Press.

Economic Policy Institute. 2006. "Hundreds of Economists Say: Raise the Minimum Wage." Retrieved from http://www.epinet.org/content.cfm/minwagestmt2006

———. 2007. EPI Issue Guide: The Minimum Wage. Retrieved from http://www.epinet.org/content.cfm/issueguides_minwage

Economy, Elizabeth. 2005. *The River Runs Black: The Environmental Challenge to China's Future*. Ithaca, NY: Cornell University Press.

Edmonds, Eric V. 2005. "Does Child Labor Decline With Improving Economic Status?" *Journal of Human Resources*, vol. 40: 77–99. University of Wisconsin Press.

Edwards, Richard. 1979. *Contested Terrain: The Transformation of the Workforce in the Twentieth Century*. New York: Basic Books.

Ehrenreich, Barbara. 2001. *Nickel and Dimed: On (Not) Getting By in America*. New York: Metropolitan.

Eisenbrey, Ross, and Ross Bernstein. 2003. "Eliminating the Right to Overtime Pay: Department of Labor Proposal Means Lower Pay, Longer Hours for Millions of Workers." Washington DC: Economic Policy Institute.

Elder, Glen. 1998. "The Life Course and Human Development." Pp. 939–91 in *Handbook of Child Psychology, vol. 1: Theoretical Models of Human Development*, edited by Richard Lerner. New York: Wiley.

———. 1999. *Children of the Great Depression: Social Change in Life Experience*. Boulder, CO: Westview Press.

Elger, Tony, and Chris Smith. 1994. *Global Japanization: The Transnational Transformation of the Labour Process*. London: Routledge.

Elliott, James, and Ryan Smith. 2001. "Ethnic Matching of Supervisors to Subordinate Work Groups: Findings on 'Bottom-up' Ascription and Social Closure." *Social Problems* 48: 258–276.

———. 2004. "Race, Gender and Workplace Power." *American Sociological Review* 69: 365–386.

Engels, Friedrich. 1936 [1845]. *The Condition of the Working-Class in England in 1844*. New York: Allen & Unwin.

England, Paula, Michelle Budig, and Nancy Folbre. 2002. "Wages of Virtue: The Relative Pay of Care Work." *Social Problems* 49: 455–473.

Epstein, David. 2005. "Fired Up at NYU." *Inside Higher Ed*. Retrieved from http://insidehighered.com/news/2005/07/13/nyu

Erikson, Kai. 1994. *A New Species of Trouble*. New York: Norton.

Fagnani, Jeanne, and Marie-Therese Letablier. 2004. "Work and Family Life Balance: The Impact of the 35 Hour Laws in France." *Work, Employment and Society* 18: 551–572.

Fantasia, Rick, and Kim Voss. 2004. *Hard Work: Remaking the American Labor Movement*. Berkeley: University of California Press.

Farnesworth-Riche, Martha. 2006. "Demographic Implications for Work-Family Research." Pp. 125–140 in *The Work and Family Handbook: Multi-Disciplinary Perspectives and Methods,* edited by Marcie Pitt-Catsouphes, Ellen Ernst Kossek, and Stephen Sweet. Mahwah, NJ: Erlbaum.

Feagin, Joe. 1991. "The Continuing Significance of Race: Antiblack Discrimination in Public Places." *American Sociological Review* 56: 101–116.

Feagin, Joe, and Robert Parker. 1990. *Building American Cities: The Urban Real Estate Game.* Englewood Cliffs, NJ: Prentice Hall.

Fernandez, Roberto, and Isabel Fernandez-Mateo. 2006. "Networks, Race and Hiring." *American Sociological Review* 71: 42–71.

Fernandez, Roberto, and Celina Su. 2004. "Space in the Study of Labor Markets." *Annual Review of Sociology* 30: 545–569.

Fine, Janice. 2006. *Worker Centers: Organizing Communities at the Edge of the Dream.* Ithaca, NY: Cornell University Press.

Fletcher, Joyce. 2001. *Disappearing Acts: Gender, Power, and Relational Practice at Work.* Cambridge, MA: MIT Press.

Florida, Richard. 2002. *The Rise of the Creative Class: And How It's Transforming Work, Leisure, Community, and Everyday Life.* New York: Basic Books.

Fredrickson, George. 1981. *White Supremacy: A Comparative Study in American and South African History.* New York: Oxford University Press.

Freeman, Richard. 2005. "Fighting for Other Folks' Wages: The Logic and Illogic of Living Wage Campaigns." *Industrial Relations* 44: 14–31.

Freud, Sigmund. 1961 [1929]. *Civilization and Its Discontents.* New York: Norton.

Freund, Peter, and George Martin. 2000. "Driving South: The Globalization of Auto Consumption and Its Social Organization of Space." *Capitalism, Nature, Socialism* 11: 51–71.

Fried, Mindy. 1998. *Taking Time: Parental Leave Policy and Corporate Culture.* Philadelphia: Temple University Press.

Friedan, Betty. 1963. *The Feminine Mystique.* New York: Norton.

Friedman, Thomas. 2005. *The World is Flat: A Brief History of the Twenty-First Century.* New York: Farrar, Straus and Giroux.

Fröbel, Folker, Jürgen Heinrichs, and Otto Kreye. 1982. *The New International Division of Labour: Structural Unemployment in Industrialised Countries and Industrialisation in Developing Countries.* New York: Cambridge University Press.

Galbraith, John Kenneth. 1976. *The Affluent Society,* 3rd ed. Boston: Houghton Mifflin.

Galinsky, Ellen, and James T. Bond. 1998. *The National Study of the Changing Work Force.* New York: Families and Work Institute.

Gallant, Mary J., and Jay E. Cross. 1993. "Wayward Puritans in the Ivory Tower: Collective Aspects of Gender Discrimination in Academia." *Sociological Quarterly* 34: 237–256.

Galtry, Judith. 2000. "Extending the 'Bright Line': Feminism, Breastfeeding, and the Workplace in the United States." *Gender & Society* 14: 295–317.

———. 2002. "Child Health: An Underplayed Variable in Parental Leave Policy Debates?" *Community, Work and Family* 5: 257–278.

Garey, Anita Ilta. 1999. *Weaving Work and Motherhood*. Philadelphia: Temple University Press.

Garfinkel, Harold. 1967. *Studies in Ethnomethodology*. Englewood Cliffs, NJ: Prentice Hall.

Gereffi, Gary. 1994. "Capitalism, Development and Global Commodity Chains." Pp. 211–231 in *Capitalism and Development*, edited by Gary Gereffi. London: Routledge.

Gergen, Kenneth. 1991. *The Saturated Self: Dilemmas of Identity in Contemporary Life*. New York: Basic Books.

Gerson, Kathleen. 2001. "Children of the Gender Revolution: Some Theoretical Questions and Findings From the Field." In *Restructuring Work and the Life Course*, edited by Victor W. Marshall, Walter R. Heinz, Helga Krueger, and Anil Verma. Toronto: University of Toronto Press.

Gerstel, Naomi, and Katherine McGonanagle. 1999. "Job Leaves and the Limits of the Family and Medical Leave Act." *Work and Occupations* 26: 510–534.

Gerstel, Naomi, and Natalia Sarkisian. 2006. "Sociological Perspectives on Families and Work: The Import of Gender, Class, and Race." Pp. 73–98 in *The Work and Family Handbook: Multi-Disciplinary Perspectives, Methods, and Approaches*, edited by Marcie Pitt-Catsouphes, Ellen Ernst Kossek, and Stephen Sweet. Boston: Erlbaum.

Gilbert, Dennis. 2003. *The American Class Structure in an Age of Growing Inequality*. Belmont, CA: Wadsworth.

Giuffre, Patti, and Christine Williams. 1994. "Boundary Lines: Labeling Sexual Harassment in Restauraunts." *Gender & Society* 8: 378–401.

Glenn, Evelyn Nakano. 2002. *Unequal Freedom: How Race and Gender Shaped American Citizenship and Labor*. Cambridge, MA: Harvard University Press.

Glover, Katherine. 2003. "No Sweat in Minneapolis." *Dollars and Sense* 249: 8–9.

Glyn, Andrew. 2005. "The Imbalance of the Global Economy." *New Left Review* 34: 5–37.

Goldberg, Jeffrey 2007. "Selling Wal-Mart: Annals of Spin." *New Yorker*, pp. 32–40.

Golden, Lonnie. 2005. "Overemployment in the US: Which Workers Face Downward Constrained Hours." In *Decent Working Time: New Trends, New Issues*, edited by Y. Boulin, J. Lallement, J. Messenger, and F. Michon. New York: International Labor Organization.

Goldfield, Mike. 1987. *The Decline of Organized Labor in the United States*. Chicago: University of Chicago Press.

Gordon, David. 1996. *Fat and Mean: The Corporate Squeeze of Working Americans and the Myth of Managerial Downsizing*. New York: Free Press.

Gordon, Jennifer. 2005. *Suburban Sweatshops: The Fight for Immigrant Rights*. Cambridge, MA: Harvard University Press.

Gornick, J. C., and M. K. Meyers. 2003. *Families That Work: Policies for Reconciling Parenthood and Employment*. New York: Russell Sage Foundation.

Gorz, Andre. 1982. *Farewell to the Working Class*. Boston: South End Press.

Graff, John de, David Wann, and Thomas Naylor. 2001. *Affluenza: The All-Consuming Epidemic*. San Francisco: Berrett-Koehler.

Granovetter, Mark. 1973. "The Strength of Weak Ties." *American Journal of Sociology* 78: 1360–1380.

———. 1995. *Getting a Job*. Chicago: University of Chicago Press.

Gratton, Brian, and Jon Moen. 2004. "Immigration, Culture, and Child Labor in the United States, 1880–1920." *Journal of Interdisciplinary History* 34: 355–391.

Green, Francis. 2006. *Demanding Work: The Paradox of Job Quality in the Affluent Economy*. Princeton, NJ: Princeton University Press.

Grunberg, Leon, Richard Anderson-Connolly, and Edward S. Greenberg. 2000. "Surviving Layoffs: The Effects on Organizational Commitment and Job Performance." *Work and Occupations* 27: 7–31.

Grusky, David, and Maria Charles. 2004. *Occupational Ghettos*, edited by David Grusky and Maria Charles. Stanford, CA: Stanford University Press.

Gusfield, Joseph. 1963. *Symbolic Crusade: Status Politics and the Urban Temperance Movement*. Chicago: University of Illinois.

Hacker, Jacob. 2006. *The Great Risk Shift: The Assault on American Jobs, Families, Health Care and Retirement and How You Can Fight Back*. New York: Oxford University Press.

Hakim, Catherine. 2001. *Work-Lifestyle Choices in the Twenty-First Century*. New York: Oxford University Press.

Hamilton, V. Lee, Clifford L. Broman, William S. Hoffman, and Deborah S. Renner. 1990. "Hard Times and Vulnerable People: Initial Effects of Plant Closing on Autoworkers' Mental Health." *Journal of Health and Social Behavior* 31: 123–140.

Harden, Jeni. 2001. "Mother Russia at Work: Gender Divisions in the Medical Profession." *European Journal of Women's Studies* 8: 181–199.

Hareven, Tamara. 1982. *Family Time and Industrial Time: The Relationship Between the Family and Work in a New England Industrial Community*. New York: Cambridge University Press.

Hareven, Tamara, and Randolph Langenbach. 1978. *Amoskeag*. New York: Pantheon Books.

Harrington, Brad, and Jaquelyn James. 2006. "The Standards of Excellence in Work-Life Integration: From Changing Policies to Changing Organizations." In *The Work and Family Handbook: Multi-Disciplinary Perspectives, Methods, and Approaches*, edited by Marcie Pitt-Catsouphes, Ellen Ernst Kossek, and Stephen Sweet. Mahwah, NJ: Erlbaum.

Harris, William. 1982. *The Harder We Run: Black Workers Since the Civil War*. New York: Oxford University Press.

Harrison, Bennett. 1997. *Lean and Mean: Why Large Corporations Will Continue to Dominate the Global Economy*. New York: Guilford Press.

Harvey, Adia. 2005. "Becoming Entrepreneurs: Intersections of Race, Class and Gender at the Black Beauty Salon." *Gender & Society* 19: 789–808.

Haugen, Steven. 2003. "Characteristics of Minimum Wage Workers in 2002." *Monthly Labor Review* 126: 37–40.

Hays, Sharon. 2003. *Flat Broke With Children: Women in the Age of Welfare Reform*. Cambridge, UK: Oxford University Press.

Head, Simon. 2003. *The New Ruthless Economy: Work and Power in the Digital Age.* Cambridge: Oxford University Press.

———. 2004. "Inside the Leviathan." *New York Review of Books* (December 16), pp. 80–89.

Heckscher, Charles. 1988. *The New Unionism: Employee Involvement in the Changing Corporation.* New York: Basic Books.

———. 1996. *White-Collar Blues: Management Loyalties in the Age of Corporate Restructuring.* New York: Basic Books.

Henson, Kevin. 1996. *Just a Temp.* Philadelphia: Temple University Press.

Henson, Kevin, and Jackie Krasas Rogers. 2001. "Why Marcia You've Changed! Male Temporary Clerical Workers Doing Masculinity in a Feminized Occupation." *Gender & Society* 15: 218–238.

Herrnstein, Richard, and Charles Murray. 1994. *The Bell Curve: Intelligence and Class Structure in American Life.* New York: Free Press.

Hessler, Peter 2003. "Underwater." *New Yorker,* July 7, pp. 28.

Heymann, Jody. 2000. *The Widening Gap: Why America's Working Families Are in Jeopardy and What Can Be Done About It.* New York: Perseus.

Hirst, Paul, and J. Zeitlin. 1991. "Flexible Specialization Versus Post-Fordism." *Economy and Society* 20: 1–56.

Hobson, Barbara. 2002. *Making Men Into Fathers: Men, Masculinities and the Social Politics of Fatherhood.* Cambridge, UK: Cambridge University Press.

Hochschild, Arlie Russell. 1983. *The Managed Heart: Commercialization of Human Feeling.* Berkeley: University of California Press.

———. 1997. *The Time Bind: When Work Becomes Home and Home Becomes Work.* New York: Metropolitan Books.

Hochschild, Arlie Russell, and Anne Machung. 1989. *The Second Shift: Working Parents and the Revolution at Home.* New York: Viking.

Hoffman, William S., Patricia Carpentier-Alting, Duane Thomas, V. Lee Hamilton, and Clifford L. Broman. 1991. "Initial Impact of Plant Closings on Automobile Workers and Their Families." *Families in Society* 72: 103–107.

Hoffnung, Michele. 2004. "Wanting It All: Career, Marriage, and Motherhood During College-Educated Women's 20s." *Sex Roles* 50: 711–723.

Holzer, Harry. 1991. "The Spatial Mismatch Hypothesis: What Has the Evidence Shown." *Urban Studies* 28: 105–122.

Honey, Michael. 1999. *Black Workers Remember: An Oral History of Segregation, Unionism, and the Freedom Struggle.* Berkeley: University of California Press.

Hostetler, Andrew, Stephen Sweet, and Phyllis Moen. 2007. "Gendered Career Paths: A Life Course Perspective on the Return to School." *Sex Roles* 56: 85–103.

Hounshell, David. 1984. *From the American System to Mass Production.* Baltimore: Johns Hopkins University Press.

Hutchens, Robert M., and Emma Dentinger. 2003. "Moving Toward Retirement." In *It's About Time: Couples and Careers,* edited by Phyllis Moen. Ithaca, NY: Cornell University Press.

Hyde, Jane Shibley, Elizabeth Fennema, and Susan Lamon. 1990. "Gender Differences in Mathematics Performance: A Meta-Analysis." *Psychological Bulletin* 107: 139–155.

Ibarra, Hermania. 1993. "Personal Networks of Women and Minorities in Management: A Conceptual Framework." *Academy of Management Review* 18: 56–87.

Iversen, Lars, and Svend Sabroe. 1988. "Psychological Well-Being Among Unemployed and Employed People After a Company Closedown: A Longitudinal Study." *Journal of Social Issues* 44: 141–152.

Iwanow, H., M. G. McEachern, and A. Jeffrey. 2005. "The Influence of Ethical Trading Policies on Consumer Apparel Purchase Decisions." *International Journal of Retail and Distribution Management:* 371–387.

Jackall, Robert. 1989. *Moral Mazes: The World of Corporate Managers.* Cambridge, UK: Oxford University Press.

Jacobs, Jerry. 1999. "The Sex Segregation of Occupations." In *Handbook of Gender and Work,* edited by Gary Powell. Thousand Oaks, CA: Sage.

Jacobs, Jerry A. and Kathleen Gerson. 2004. *The Time Divide: Work, Family, and Gender Inequality.* Cambridge, MA: Harvard University Press.

Jacoby, Sanford. 1985. *Employing Bureaucracy.* New York: Columbia University Press.

———. 1991. *Masters to Managers.* New York: Columbia University Press.

———. 1999. "Are Career Jobs Headed for Extinction?" *California Management Review* 42: 123–145.

———. 2001. "Risk and the Labor Market: Societal Past as Economic Prologue." In *Sourcebook of Labor Markets: Evolving Structures and Processes,* edited by Ivar Berg and Arne Kalleberg. New York: Kluwer.

Jahoda, Marie. 1982. *Employment and Unemployment: A Social Psychological Analysis.* New York: Cambridge University Press.

Jencks, Christopher, and Paul Peterson. 1991. *The Urban Underclass.* Washington, DC: Brookings Institution.

Johnston, David Cay. 2004. *Perfectly Legal: The Covert Campaign to Rig Our Tax System to Benefit the Super Rich—and Cheat Everybody Else.* New York: Portfolio.

Jones, Ellis, Ross Haenfler, Brett Johnson, and Brian Klocke. 2002. *The Better World Handbook: From Good Intentions to Everyday Actions.* Gabriola Island, BC: New Society.

Jones, Marc T. 2005. "The Transnational Corporation, Corporate Social Responsibility and the 'Outsourcing' Debate." *Journal of American Academy of Business, Cambridge* 6: 91–97.

Kahlenberg, Richard. 2004. *America's Untapped Resource: Low-Income Students in Higher Education.* New York: Century Foundation Press.

Kahn, Joseph. 2003. "Ruse in Toyland: Chinese Workers' Hidden Woe." *New York Times* (December 7), p. 2.

Kalleberg, Arne. 2007. *The Mismatched Worker.* New York: Norton.

Kalleberg, Arne, Barbara Reskin, and Ken Hudson. 2000. "Bad Jobs in America: Standard and Nonstandard Employment Relations and Job Quality in the United States." *American Sociological Review* 65: 256–278.

Kamerman, Sheila. 2000. "Parental Leave Policies: An Essential Ingredient in Early Childhood Education and Care Policies." Ann Arbor, MI: Society for Research in Child Development, University of Michigan.

Kanter, Rosabeth Moss. 1977. *Men and Women of the Corporation.* New York: Basic Books.

Karlin, Carolyn Aman, Paula England, and Mary Richardson. 2002. "Why Do 'Women's Jobs' Have Low Pay for Their Educational Level?" *Gender Issues* 20: 3–22.

Katz, Michael. 1996. *In the Shadow of the Poorhouse.* New York: Basic Books.

Kelly, Erin. 2006. "Work-Family Policies: The United States in International Perspective." Pp. 99–124 in *The Work and Family Handbook: Multi-Disciplinary Perspectives, Methods, and Approaches,* edited by Marcie Pitt-Catsouphes, Ellen Ernst Kossek, and Stephen Sweet. Boston: Erlbaum.

Kelvin, P., and J. E. Jarrett. 1985. *Unemployment: Its Social Psychological Effects.* Cambridge, UK: Cambridge University Press.

Kenessey, Zoltan. 1987. "Primary, Secondary, Tertiary and Quarternary Sectors of the Economy." *Review of Income and Wealth* 33: 359–386.

Kennelly, Ivy. 1999. "'That Single Mother Element': How White Employers Typify Black Women." *Gender & Society* 13: 168–192.

———. 2002. "'I Would Never Be A Secretary': Reinforcing Gender in Segregated and Integrated Occupations." *Gender & Society* 16: 603–624.

———. 2006. "Secretarial Work, Nurturing, and the Ethic of Service." *NWSA Journal* 18: 170–192.

Kets de Vries, Manfred, and Katarina Balazs. 1997. "The Downside of Downsizing." *Human Relations* 50: 11–50.

Khadria, Binod. 2001. "Shifting Paradigms of Globalization: The Twenty-First Century Transition Towards Generics in Skilled Migration from India." *International Migration,* 39: 45.

Kim, Chigon, and Mark Gottdiener. 2004. "Urban Problems in Global Perspective." Pp. 172–192 in *Handbook of Social Problems: A Comparative International Perspective,* edited by George Ritzer. Thousand Oaks, CA: Sage.

Kimmel, Michael. 2006. "A War Against Boys?" *Dissent* 53: 65–70.

Kirschenmann, Jolene, and Kathryn Neckerman. 1991. "'We'd Love to Hire Them But . . . ': The Meaning of Race for Employers." Pp. 203–232 in *The Urban Underclass,* edited by Christopher Jencks and Paul Peterson. Washington DC: Brookings Institution.

Kochan, Thomas. 2005. *Restoring the American Dream: A Working Families' Agenda for America.* Cambridge, MA: MIT Press.

Kochan, Thomas, Harry Katz, and Robert McKersie. 1986. *The Transformation of American Industrial Relations.* New York: Basic Books.

Koeber, Charles. 2002. "Corporate Restructuring, Downsizing and the Middle Class: The Process and Meaning of Worker Displacement in the 'New Economy.'" *Qualitative Sociology* 25: 217–246.

Kossek, Ellen Ernst, and Alyssa Fried. 2006. "The Business Case: Managerial Perspectives on Work and Family." In *The Work and Family Handbook: Multi-Disciplinary Perspectives, Methods, and Approaches,* edited by Marcie Pitt-Catsouphes, Ellen Ernst Kossek, and Stephen Sweet. Boston: Erlbaum.

Kossek, Ellen Ernst, Brenda Lautsch, and Susan Eaton. 2006. "Telecommuting, Control, and Boundary Management: Correlates of Policy Use and Practice, Job Control, and Work-Family Effectiveness." *Journal of Vocational Behavior* 68: 347–367.

Kossek, Ellen Ernst, and Mary Dean Lee. 2005. *Benchmarking Survey: A Snapshot of Human Resource Managers' Perspectives on Implementing Reduced-Load Work for Professionals.* East Lansing: Michigan State University, Montreal: McGill University. Retrieved from http://www.lir.msu.edu/workingpapers/documents/benchmark.pdf

Kossek, Ellen Ernst, and Cynthia Ozeki. 1999. "Bridging the Work-Family Policy and Productivity Gap." *Community, Work & Family* 2: 7–32.

Kozol, Jonathan. 2006. *The Shame of the Nation.* New York: Three Rivers Press.

Kremer, Monique. 2006a. "Consumers in Charge of Care: The Dutch Personal Budget and Its Impact on the Market, Professionals, and the Family." *European Societies* 8: 385–401.

———. 2006b. "The Politics of Ideals of Care: Danish and Flemish Child Care Policy Compared." *Social Politics* 13: 261–285.

Krugman, Paul. 2005. "French Family Values." *New York Times* (July 29), p. 1.

Kurlansky, Mark. 1998. *Cod: A Biography of the Fish That Changed the World.* New York: Penguin.

Lafer, Gordon. 2003. "Graduate Student Unions: Organizing in a Changed Academic Economy." *Labor Studies Journal* 28: 25–43.

Lambert, Susan. 2003. "Added Benefits: The Link Between Work Life Benefits and Organizational Citizenship Behavior." *Academy of Management Journal* 43: 801–815.

Lasch, Christopher. 1995. *Haven in a Heartless World: The Family Besieged.* New York: Norton.

Lash, Scott, and John Urry. 1987. *The End of Organized Capitalism.* Madison: University of Wisconsin Press.

Latack, Janina C., Angelo Kinicki, and Gregory E. Prussia. 1995. "An Integrative Process Model of Coping with Job Loss." *Academy of Management Review* 20(2): 311–334.

Leana, Carrie R., and Daniel C. Feldman. 1992. *Coping With Job Loss: How Individuals, Organizations, and Communities Respond to Layoffs.* New York: Lexington Books.

———. 1995. "Finding New Jobs after a Plant Closing: Antecedents and Outcomes of the Occurrence and Quality of Reemployment." *Human Relations* 48: 1381–1401.

Lee, Ching Kwan. 1998. *Gender and the South China Miracle.* Berkeley: University of California Press.

Levy, Frank. 1998. *The New Dollars and Dreams.* New York: Russell Sage Foundation.

Lichtenstein, Nelson. 2002. *State of the Union: A Century of American Labor.* Princeton, NJ: Princeton University Press.

Liebow, Eliot. 1967. *Tally's Corner*. Lanham, MD: Rowman & LIttlefield.

Lim, Nelson. 2001. "On the Backs of Blacks? Immigrants and the Fortunes of African Americans." Pp. 186–227 in *Strangers at the Gates: New Immigrants in Urban America*, edited by Roger Waldinger. Berkeley: University of California Press.

Lipset, Seymour. 1996. *American Exceptionalism: A Double-Edged Sword*. New York: Norton.

Lopez, Steven. 2004. "Overcoming Legacies of Business Unionism: Why Grassroots Organizing Activities Succeed." Pp. 114–132 in *Rebuilding Labor: Organizers and Organizing in the New Labor Movement*, edited by Ruth Milkman and Kim Voss. Ithaca, NY: ILR Press.

Lorber, Judith. 1994. *Paradoxes of Gender*. New Haven, CT: Yale University Press.

Lott, Bernice, and Diane Maluso. 1993. "The Social Learning of Gender." In *The Psychology of Gender*, edited by Anne Beall and Ellen Berscheid. New York: Guilford Press.

Lowenstein, Roger. 2006. "Do Illegal Immigrants Take Jobs?" *New York Times Magazine* (July 9), pp. 34–43, 60–71.

Luken, Ralph, and Rodney Stares. 2005. "Small Business Responsibility in Developing Countries: A Threat or an Opportunity?" *Business Strategy and the Environment* 14: 38–53.

Machlowitz, M. M. 1980. *Workaholics: Living with Them, Working with Them*. Reading, MA: Addison-Wesley.

MacKinnon, Catherine. 1998. *Sex Equality*. New York: Foundation Press.

MacLeod, Jay. 1995. *Ain't No Makin' It*. Boulder, CO: Westview Press.

MacMillan, Ross. 2005. "The Structure of the Life Course: Standardized? Individualized? Differentiated?" in *Advances in Life Course Research*, vol. 9. New York: Elsevier.

Mandel, Hadas, and Moshe Semyonov. 2004. "Family Policies, Wage Structures, and Gender Gaps: Sources of Earnings Inequality in 20 Countries." *American Sociological Review* 70: 949–967.

———. 2006. "A Welfare State Paradox: State Interventions and Women's Employment Opportunities in 22 Countries." *American Journal of Sociology* 111: 1910–1949.

Marglin, Steven. 1982. "What Do Bosses Do?" In *Classes, Power, and Conflict: Classical and Contemporary Debates*, edited by Anthony Giddens and David Held. Berkeley: University of California Press.

Marks, Alexandra. 2004. "United's Pension Woes: Sign of a Bigger Issue." *Christian Science Monitor* (October 4), p. 2.

Marshall, Ray, and Marc Tucker. 1992. *Thinking for a Living: Education and the Wealth of Nations*. New York: Basic Books.

Martin, Susan Ehrlich. 1982. *Breaking and Entering: Policewomen on Patrol*. Los Angeles: University of California Press.

Marx, Karl. 1964 [1844]. "Economic and Philosophic Manuscripts." In *Karl Marx: Economic and Philosophic Manuscripts*, edited by Thomas Bottomore. New York: McGraw Hill.

———. 1970 [1867]. *Capital*. London: Lawrence and Wishart.

Marx, Karl, and Friedrich Engels. 1972 [1848]. "Manifesto of the Communist Party." In *The Marx-Engels Reader,* edited by Robert C. Tucker. New York: Norton.

Masser, Barbara, and Dominic Abrams. 2004. "Reinforcing the Glass Ceiling: The Consequences of Hostile Sexism for Female Managerial Candidates." *Sex Roles* 51: 609–615.

Massey, Douglas. 2006. "The Wall That Keeps Illegal Workers In." *New York Times* (April 4), pp. 8–12.

Massey, Douglas, and Katherine Bartley. 2005. "The Changing Legal Status Distribution of Immigrants: A Caution." *International Migration Review* 39: 469–85.

Mattingly, Doreen. 1999. "Making Maids." Pp. 61–78 in *Gender, Migration and Domestic Service,* edited by Doreen Mattingly and Janet Momsen. New York: Routledge.

Mattsson, Lars-Gunnar. 2003. "Reorganization of Distribution in Globalization of Markets: The Dynamic Context of Supply Chain Management." *Supply Chain Management* 8: 416–426.

McCann, Deirdre. 2005. "Working Time Laws: A Global Perspective: Findings From the ILO's Conditions of Work and Employment Database." Geneva, Switzerland: International Labor Office.

McElhinney, Stephen. 2005. "Exposing the Interests: Decoding the Promise of the Global Knowledge Society." *New Media & Society* 7: 748–769.

McPhee, John. 2005. "Out in the Sort; Annals of Transport." *New Yorker* 81: 161–167.

Meiksins, Peter. 1998. "The Time Bind." *Monthly Review* 49:1–13.

Meiksins, Peter, and Peter Whalley. 2002. *Putting Work in Its Place: A Quiet Revolution,* Ithaca, NY: ILR Press.

———. 2004. "Labor and Leisure: Should Europe Work More, or America Less?" *International Herald Tribune* (Houston, TX, August 11), p. 4.

Meyer, Steven. 1981. *The Five Dollar Day: Labor Management and Social Control in the Ford Motor Company, 1908–1921.* Albany: State University of New York Press.

Milberg, William. 2004. "The Changing Structure of Trade Linked to Global Production Systems: What Are the Policy Implications?" *International Labour Review* 143: 45–90.

Milkman, Ruth. 1997. *Farewell to the Factory: Auto Workers in the Late Twentieth Century.* Berkeley: University of California Press.

———. 2006. *L.A. Story: Immigrant Workers and the Future of the U.S. Labor Movement.* Berkeley: University of California Press.

Milkman, Ruth, and Kim Voss. 2004. *Rebuilding Labor: Organizers and Organizing in the New Labor Movement.* Ithaca, NY: ILR Press.

Mills, C. Wright. 1959. *The Sociological Imagination.* New York: Oxford University Press.

———. 1978. *The Power Elite.* New York: Oxford University Press.

———. 2002 [1951]. *White Collar: The American Middle Classes.* London: Oxford University Press.

Mir, Ali, Biju Mathew, and Raza Mir. 2000. "The Codes of Migration: Contours of the Global Software Labor Market." *Cultural Dynamics* 12: 5.

Mishel, Lawrence, Jared Bernstein, and Sylvia Allegretto. 2005. *The State of Working America 2004/2005.* Ithaca, NY: Economic Policy Institute.

Mishra, A. K., and G. M. Spreitzer. 1998. "Explaining How Survivors Respond to Downsizing." *Academy of Management Review* 23: 567–588.

MIT. 1999. "A Study on the Status of Women Faculty in Science.at MIT." Boston: Massachusetts Institute of Technology. Retrieved from http://web.mit.edu/fnl/women/women.html

Moberg, David. 2005. "Look Who's Walking." vol. 2006: The Nation Online. Retrieved from http://www.thenation.com/doc/20050801/moberg

Moen, Phyllis. 2001a. "Constructing a Life Course." *Marriage and Family Review* 30: 97–109.

———. 2001b. "The Gendered Life Course." In *Handbook of Aging and the Social Sciences,* edited by L. George and R. H. Binstock. San Diego: Academic Press.

———. 2003. *It's About Time: Couples and Careers.* Ithaca: Cornell University Press.

Moen, Phyllis, and Patricia V. Roehling. 2005. *The Career Mystique.* Boulder, CO: Rowman & Littlefield.

Moen, Phyllis, and Stephen Sweet. 2003. "Time Clocks: Couples' Work Hour Strategies." Pp. 17–34 in *It's About Time: Career Strains, Strategies, and Successes,* edited by Phyllis Moen. Ithaca, NY: Cornell University Press.

———. 2004. "From 'Work-Family' to 'Flexible Careers': A Life Course Reframing." *Community, Work and Family* 7: 209–226.

Moen, Phyllis, Stephen Sweet, and Raymond Swisher. 2005. "Embedded Career Clocks: The Case of Retirement Planning." Pp. 237–268 in *The Structure of the Life Course: Standardized? Individualized? Differentiated?* vol. 9, edited by Ross MacMillan. New York: Elsevier.

Moen, Phyllis, Stephen Sweet, and Bickley Townsend. 2001. "How Family Friendly is Upstate New York?" Ithaca, NY: Cornell Careers Institute.

Moen, Phyllis, Ronit Waismel-Manor, and Stephen Sweet. 2003. "Success." Pp. 133–152 in *It's About Time: Couples and Careers,* edited by Phyllis Moen. Ithaca, NY: Cornell University Press.

Moen, Phyllis, and E. Wethington. 1992. "The Concept of Family Adaptive Strategies." *Annual Review of Sociology* 18: 233–251.

Mohanty, Chandra Talpade. 2003. *Feminism Without Borders: Decolonizing Theory, Practicing Solidarity.* Durham, NC: Duke University Press.

Montgomery, David. 1979. *Workers' Control in America: Studies in the History of Work, Technology, and Labor Struggles.* Cambridge, UK: Cambridge University Press.

Moody, Kim. 1997. *Workers in a Lean World.* New York: Verso.

Moore, Thomas. 1996. *The Disposable Work Force: Worker Displacement and Employment Instability in America.* New York: Aldine De Gruyter.

Morgan, Kathleen O'Leary, and Scott Morgan. 2007. *State Rankings 2007: A Statistical View of the United States.* Lawrence, KS: Morgan Quinto Press.

Morgan, Kimberly. 2005. "The "Production" of Child Care: How Labor Markets Shape Social Policy and Vice Versa." *Social Politics* 12: 243–263.

Morgan, Kimberly, and Kathrin Zippel. 2003. "Paid to Care: The Origins and Effects of Care Leave Policies in Western Europe." *Social Politics* 10: 49–85.

Moss, Philip, and Chris Tilly. 2001. *Stories Employers Tell: Race, Skill, and Hiring in America.* New York: Russell Sage Foundation.

Mouw, Ted. 2002. "Are Black Workers Missing the Connection? The Effect of Spatial Distance and Employee Referrals on Interfirm Racial Segregation." *Demography* 39: 507–528.

Murray, Charles. 1995. *Losing Ground: American Social Policy, 1950–1980.* New York: Basic Books.

Myrdal, Gunnar. 1995 [1954]. *An American Dilemma: The Negro Problem and Modern Democracy.* New York: Transaction.

National Institute for Occupational Safety and Health. 2007. "Stress at Work." Publication No. 99:101: National Institute for Occupational Safety and Health. Retrieved from http://www.cdc.gov/niosh/stresswk.html

Naumann, Stefanie E., Nathan Bennett, Robert J. Bies, Christopher L. Martin. 1998. "Laid Off, but Still Loyal." *International Journal of Conflict Management* 9: 356–368.

Neal, Margaret, and Leslie Hammer. 2006. *Working Couples Caring for Children and Aging Parents: Effects on Work and Well-Being.* Mahwah, NJ: Erlbaum.

Nee, Victor, Jimy Sanders, and Scott Sernau. 1994. "Job Transitions in an Immigrant Metropolis: Ethnic Boundaries and the Mixed Economy." *American Sociological Review* 59: 849–872.

Negrey, Cynthia. 1993. *Gender, Time, and Reduced Work.* Albany: State University of New York Press.

Nelson, Daniel. 1980. *Frederick W. Taylor and the Rise of Scientific Management.* Madison: University of Wisconsin Press.

Nelson, Valerie, Adrienne Martin, and Joachim Ewert. 2005. "What Difference Can They Make? Assessing the Social Impact of Corporate Codes of Practice." *Development in Practice* 15: 539–545.

Nestle, Marion. 2003. *Food Politics.* Los Angeles: University of California Press.

Neumark, David. 2000. *On the Job: Is Long-Term Employment a Thing of the Past?* New York: Russell Sage Foundation.

Newman, Katherine. 1999. *No Shame in My Game: The Working Poor in the Inner City.* New York: Knopf and Russell Sage Foundation.

———. 2006. *Chutes and Ladders.* New York: Russell Sage Foundation.

Ngai, Pun. 2005. *Made in China: Women Factory Workers in a Global Workplace.* Durham, NC: Duke University Press.

Nippert-Eng, Christine. 1996. *Home and Work.* Chicago: University of Chicago Press.

Noble, David. 1979. *America by Design: Science, Technology, and the Rise of Corporate Capitalism.* New York: Knopf.

———. 1984. *Forces of Production.* New York: Knopf.

Nowicki, Carol. 2003. "Family and Medical Leave Act." Sloan Work and Family Encyclopedia. Retrieved from http://wfnetwork.bc.edu/encyclopedia_template .php?id=234

O'Brien, Helena, and Sarita Gupta. 2005. "Local Power Can Change Wal-Mart: The ACORN and Jobs With Justice Organizing Strategy." Social Policy 36: 16–20.

Ochs, Elinor, Anthony Graesch, Angela Mittmann, Thomas Bradbury, and Rena Repetti. 2006. "Video Ethnography and Ethnoarcheaological Tracking." Pp. 387–410 in The Work and Family Handbook: Multi-Disciplinary Perspectives, Methods, and Approaches, edited by Marcie Pitt-Catsouphes, Ellen Ernst Kossek, and Stephen Sweet. Mahwah, NJ: Erlbaum.

O'Leary, Christopher and Stephen Wandner. 1997. "Summing Up: Achievements, Problems and Prospects." In Unemployment Insurance in the United States: Analysis of Policy Issues, edited by Christopher O'Leary and Stephen Wandner. New York: W. E. Upjohn Institute.

Omi, Michael, and Howard Winant. 1986. Racial Formation in the United States: From the 1960s to the 1980s. New York: Routledge.

O'Reilly, Brian 1994. "The New Deal: What Companies and Employees Owe One Another." Fortune (June 13), pp. 44–50.

Organisation for Economic Co-operation and Development. 2004. Wages and Benefits: OECD Indicators. Paris: OECD.

———. 2006. OECD Factbook: Economic, Environmental and Social Statistics. Paris: OECD.

Osterman, Paul. 2001. Securing Prosperity. Princeton, NJ: Princeton University Press.

Osterud, N. G. 1987. "'She Helped Me Hay It as Good as a Man': Relations Among Women and Men in an Agricultural Community." Pp. 87–97 in "To Toil the Livelong Day": America's Women at Work, 1780–1980, edited by C. Groneman and M. B. Norton. Ithaca, NY: Cornell University Press.

Padavic, Irene, and Barbara Reskin. 2002. Women and Men at Work. Thousand Oaks, CA: Pine Forge Press.

Parker, Mike. 1985. Inside the Circle: A Union Guide to QWL. Boston: South End Press.

Parker, Mike, and Jane Slaughter. 1988. Unions and the Team Concept. Boston: South End Press.

Parthasarathy, Balaji. 2004. "India's Silicon Valley or Silicon Valley's India? Socially Embedding the Computer Software Industry in Bangalore." International Journal of Urban and Regional Research 28: 664–685.

Patni, Ambika. 1999. "Silicon Valley of the East." Harvard International Review 21: 8.

Paules, Greta Foff. 1991. Dishing it Out: Power and Resistance Among Waitresses in a New Jersey Restaurant. Philadelphia: Temple University Press.

Pavalko, Eliza, and Kathryn Henderson. 2006. "Combining Care Work and Paid Work: Do Workplace Policies Make a Difference?" Research on Aging 28: 359–374.

Pelsmacker, Patrick De, Liesbeth Driesen, and Glenn Rayp. 2005. "Do Consumers Care About Ethics? Willingness to Pay for Fair Trade Coffee." *Journal of Consumer Affairs* 39: 363–386.

Perrucci, Carolyn C. 1994. "Economic Strain, Family Structure and Problems With Children Among Displaced Workers." *Journal of Sociology and Social Welfare* 21: 79–91.

Perrucci, Carolyn C., Robert Perrucci, and Dena B. Targ. 1997. "Gender Differences in the Economic, Psychological and Social Effects of Plant Closings in an Expanding Economy." *Social Science Journal* 34: 217–233.

Perrucci, Robert, and Carl Wysong. 2002. *The New Class Society: Goodbye American Dream?* New York: Rowman & Littlefield.

Pettit, Becky, and Jennifer Hook. 2005. "The Structure of Women's Employment in Comparative Perspective." *Social Forces* 84: 779–801.

Pfau-Effinger, Birgit. 2004. *Development of Culture, Welfare States and Women's Employment in Europe.* Burlington, VT: Ashgate.

Pietrykowski, Bruce. 1999. "Beyond the Fordist/Post-Fordist Dichotomy: Working Through the Second Industrial Divide." *Review of Social Economy* 57: 177–198.

Piore, Michael. 1977. "The Dual Labor Market and its Implications." In *Problems in Political Economy*, edited by David Gordon. Lexington, MA: D. C. Heath.

Piore, Michael, and Charles Sabel. 1984. *The Second Industrial Divide: Possibilities for Prosperity.* New York: Basic Books.

Piore, Michael, and Sean Safford. 2006. "Changing Regimes of Workplace Governance: Shifting Axes of Social Mobilization and the Challenge to Industrial Relations Theory." *Industrial Relations* 45: 299–325.

Pitt-Catsouphes, Marcie, Ellen Ernst Kossek, and Stephen Sweet. 2006. *The Work and Family Handbook: Multi-Disciplinary Perspectives, Methods, and Approaches.* Mahwah, NJ: Erlbaum.

Pixley, Joy E., and Phyllis Moen. 2003. "Prioritizing Careers." Pp. 183–202 in *It's About Time: Couples and Careers*, edited by Phyllis Moen. Ithaca, NY: Cornell University Press.

Polanyi, Karl. 1944. *The Great Transformation.* New York: Farrar and Rinehart.

Polivka, Anne E., and Thomas Nardone. 1989. "On the Definition of 'Contingent Work.'" *Monthly Labor Review* 112: 9–16.

Pollert, Anna. 1988. "Dismantling Flexibility." *Capital and Class* 34: 42–75.

Portes, Alejandro, and John Walton. 1981. *Labor, Class, and the International System.* New York: Academic Press.

Portes, Alejandro, and Ruben Rumbaut. 2001. *Legacies: The Story of the Immigrant Second Generation.* Berkeley: University of California Press.

Portes, Alejandro, and Min Zhou. 1993. "The New Second Generation: Segmented Assimilation and Its Variants." *Annals of the American Academy of Political and Social Sciences* 530: 74–96.

Poster, Winifred, and Srirupa Prasad. 2005. "Work-Family Relations in Transnational Perspective: A View From High-Tech Firms in India and the United States." *Social Problems* 52: 122–146.

Presser, Harriet B. 2000. "Nonstandard Work Schedules and Marital Instability." *Journal of Marriage and the Family* 62: 93–110.

———. 2003. *Working in a 24/7 Economy: Challenges for American Families.* New York: Russell Sage Foundation.

Preston, Anne. 2004. *Leaving Science: Occupational Exit From Scientific Careers.* New York: Russell Sage Foundation.

Putnam, Robert. 2000. *Bowling Alone: The Collapse and Revival of American Community.* New York: Simon & Schuster.

Raider, Holly, and Ronald Burt. 1996. "Boundaryless Careers and Social Capital." In *The Boundaryless Career: A New Employment Principle for a New Organizational Era,* edited by Michael B. Arthur and Denise M. Rousseau. New York: Oxford University Press.

Reich, Robert. 2002. *The Future of Success: Working and Living in the New Economy.* New York: Vintage.

Reid, Peter. 1989. "How Harley Beat Back the Japanese." *Fortune* 120: 155–159.

Reskin, Barbara. 1998. *The Realities of Affirmative Action in Employment.* Washington, DC: American Sociological Association.

Reskin, Barbara, Debra McBrier, and Julie Kmec. 1999. "The Determinants and Consequences of Workplace Sex and Race Composition." *Annual Review of Sociology* 25: 335–361.

Reskin, Barbara, and Patricia Roos. 1991. *Job Queues, Gender Queues.* Philadelphia: Temple University Press.

Richardson, Pete. 2006. "The Anthropology of the Workplace and the Family." Pp. 165–188 in *The Work and Family Handbook: Multi-Disciplinary Perspectives, Methods, and Approaches,* edited by Marcie Pitt-Catsouphes, Ellen Ernst Kossek, and Stephen Sweet. Mahwah, NJ: Erlbaum.

Riesman, David, Nathan Glazer, and Reuel Denney. 2001 [1961]. *The Lonely Crowd: A Study of the Changing American Character.* New Haven, CT: Yale University Press.

Rifkin, Jeremy. 2004. *The End of Work.* New York: Jeremy P. Tarcher/Putnam Books.

Riley, Matilda White, and John W. Riley Jr. 2000. "Age Integration: Conceptual and Historical Background." *Gerontologist* 40: 266–270.

Rimer, Sara. 2007. "For Girls, It's Be Yourself, and Be Perfect, Too." *New York Times* (April 1), p. 1.

Rinehart, James, Christopher Huxley, and David Robertson. 1997. *Just Another Car Factory? Lean Production and Its Discontents.* Ithaca, NY: ILR Press.

Ritzer, George. 1996. *The McDonaldization of Society.* Thousand Oaks, CA: Pine Forge Press.

Roberts, Bryan. 2005. "Globalization and Latin American Cities." *International Journal of Urban and Regional Research* 29: 110–123.

Robinson, John, and Geoffrey Godbey. 1997. *Time for Life. The Surprising Ways Americans Use Their Free Time.* University Park: Pennsylvania State University Press.

Roehling, Patricia V., Mark V. Roehling, and Phyllis Moen. 2001. "The Relationship between Work-Life Policies and Practices and Employee Loyalty: A Life Course Perspective." *Journal of Family and Economic Issues* 22: 141–170.

Roy, D. 1955. "Efficiency and 'The Fix.'" *American Journal of Sociology* 60: 255–266.

Royster, Deirdre. 2003. *Race and the Invisible Hand: How White Networks Exclude Black Men from Blue-Collar Jobs.* Berkeley: University of California Press.

Rubin, Beth A. 1995. *Shifts in the Social Contract: Understanding Change in American Society.* Los Angeles: Pine Forge Press.

Rubin, Beth A., and Charles Brody. 2005. "Contradictions of Commitment in the New Economy: Insecurity, Time, and Technology." *Social Science Research* 34: 843–861.

Rudnyckyj, Daromir. 2004. "Technologies of Servitude: Governmentality and Indonesian Transnational Labor Migration." *Anthropological Quarterly* 77: 407–434.

Rybczynski, Witold. 1991. *Waiting for the Weekend.* New York: Viking.

Sabel, Charles, and J. Zeitlin. 1985. "Historical Alternatives to Mass Production." *Past and Present* 108: 133–175.

Sahlins, Marshall. 1972. *Stone Age Economics.* Chicago: Aldine-Atherton.

Saiz, Albert. 2003. "The Impact of Immigration on American Cities: An Introduction to the Issues." *Business Review* Q4: 4–23.

Sales, Esther. 1995. "Surviving Unemployment: Economic Resources and Job Loss Duration in Blue-Collar Households." *Social Work* 40: 483–494.

Sampson, Robert, Jeffrey D. Morenoff, and Thomas Gannon Rowley. 2002. "Assessing 'Neighborhood Effects': Social Processes and New Directions in Research." *Annual Review of Sociology* 51: 443–478.

Sampson, Robert, S. W. Raudenbush, and Felton Earls. 1997. "Neighborhoods and Violent Crime: Multilevel Study of Collective Efficacy." *Science* 277: 918–924.

Santorum, Rick. 2006. *It Takes a Family: Conservatism and the Common Good.* New York: Intercollegiate Studies Institute.

Sarkisian, Natalia, and Naomi Gerstel. 2004. "Explaining the Gender Gap in Help to Parents: The Importance of Employment." *Journal of Marriage and Family* 66: 431–451.

Sassen, Saskia. 1995. "Immigration and Local Labor Markets." Pp. 87–127 in *The Economic Sociology of Immigration: Essays on Networks, Ethnicity, and Entrepreneurship,* edited by Alejandro Portes. New York: Russell Sage Foundation.

Sassen, Saskia, and R.C. Smith. 1992. "Post-Industrial Growth and Economic Reorganization: The Impact on Immigrant Employment." In *United States-Mexico Relations: Labour Market Interdependence,* edited by J.A. Bustamante, C.W. Reynolds, and R.A. Honojosa Ojeda. Stanford, CA: Stanford University Press.

Saxenian, Anna Lee. 1996. "Beyond Boundaries: Open Labor Markets and Learning in Silicon Valley." In *The Boundaryless Career: A New Employment Principle for a New Organizational Era,* edited by Michael B. Arthur and Denise M. Rousseau. New York: Oxford University Press.

Schlosser, Eric. 2005. *Fast Food Nation*. New York: Harper Perennial.

Schmidt, Stefanie. 2000. "Job Security Beliefs in the General Social Survey: Evidence on Long-Run Trends and Comparability with Other Surveys." In *On the Job: Is Long-Term Employment a Thing of the Past?* Edited by David Neumark. New York: Russell Sage Foundation.

Schor, Juliet. 1991. *The Overworked American: The Unexpected Decline of Leisure*. New York: Basic Books.

———. 1998. *The Overspent American: Upscaling, Downshifting, and the New Consumer*. New York: Basic Books.

Schrage, Elliot. 2004. "Supply and the Brand." *Harvard Business Review* 82: 20–21.

Schrank, Andrew. 2004. "Ready-to-Wear Development? Foreign Investment, Technology Transfer, and Learning by Watching in the Apparel Trade." *Social Forces* 83: 123–156.

Schuman, Howard, Charlotte Steeh, Lawrence Bobo, and Maria Krysan. 1998. *Racial Attitudes in America: Trends and Interpretations* (rev. edition). Cambridge, MA: Harvard University Press.

Schumpeter, Joseph. 1989. *Essays: On Entrepreneurs, Innovations, Business Cycles, and the Evolution of Capitalism*. New Brunswick, NJ: Transaction.

Seidman, Gay. 2007. *Beyond the Boycott: Labor Rights, Human Rights, and Transnational Activism*. New York: Russell Sage Foundation.

Sennett, Richard. 1998. *The Corrosion of Character: The Personal Consequences of Work in the New Capitalism*. New York: Norton.

Shaiken, Harley. 1984. *Work Transformed: Automation and Labor in the Computer Age*. New York: Lexington Books.

Shamir, Boas. 1986a. "Protestant Work Ethic, Work Involvement and the Psychological Impact of Unemployment." *Journal of Occupational Behaviour* 7: 25–38.

———. 1986b. "Unemployment and Household Division of Labor." *Journal of Marriage and the Family* 48: 195–206.

Shephard, Roy, and Jean Bonneau. 2002. "Assuring Gender Equity in Recruitment Standards for Police Officers." *Canadian Journal of Applied Physiology* 27: 263–295.

Shih, Johanna. 2004. "Project Time in Silicon Valley." *Qualitative Sociology* 27: 223–245.

Shulman, Beth. 2005. *The Betrayal of Work: How Low-Wage Jobs Fail 30 Million Americans and Their Families*. New York: New Press.

Skocpol, Theda. 1992. *Protecting Soldiers and Mothers: The Political Origins of Social Policy in the United States*. Boston: Harvard University Press.

Smith, Brian T., and Beth A. Rubin. 1997. "From Displacement to Reemployment: Job Acquisition in the Flexible Economy." *Social Science Research* 26: 292–308.

Smith, Chris, and Peter Meiksins. 1995. "System Society and Dominance Effects in Cross-National Organisational Analysis." *Work, Employment and Society* 9: 241–261.

Smith, Craig. 2006. "Letter From Paris: 4 Simple Rules for Firing An Employee in France." *International Herald Tribune* (March 29), p. 1.

Smith, Ryan. 1997. "Race, Income and Authority at Work: A Cross-Temporal Analysis of Black and White Men, 1972–1994." *Social Problems* 44: 19–37.

Smith, Vicki. 1990. *Managing in the Corporate Interest: Control and Resistance in an American Bank.* Berkeley: University of California Press.

———. 2002. *Crossing the Great Divide: Worker Risk and Opportunity in the New Economy.* Ithaca, NY: Cornell University Press.

Snipp, Matthew, and Gene Summers. 1992. "American Indians and Economic Poverty." In *Rural Poverty in America,* edited by Cynthia Duncan. New York: Auburn House.

Sobel, Mechal. 1987. *The World They Made Together: Black and White Values in Eighteenth-Century Virginia.* Princeton, NJ: Princeton University Press.

Stack, Carol B. 1997 [1974]. *All Our Kin: Strategies for Survival in a Black Community.* New York: Basic Books.

Stainback, Kevin, Corre Robinson, and Donald Tomaskovic-Devey. 2005. "Race and Workplace Integration: A Politically Mediated Process?" *American Behavioral Scientist* 48: 1200–1228.

Statistical Abstracts of the United States. 2006 (and earlier years). *Statistical Abstracts of the United States.* Washington, DC: U.S. Government Printing Office. Retrieved from http://www.census.gov/compendia/statab/

Steele, Shelby. 1990. *The Content of Our Character: A New Vision of Race in America.* New York: HarperCollins.

Sullivan, Mercer. 1989. *Getting Paid: Youth Crime and Work in the Inner City.* Ithaca, NY: Cornell University Press.

Sun, Huey-Lian, and Alex P. Tang. 1998. "The Intra-Industry Information Effect of Downsizing Announcements." *American Business Review* 16(2): 68–76.

Sweet, Stephen. 2007. "The Older Worker, Job Insecurity and the New Economy." *Generations* 31: 45–49.

Sweet, Stephen, and Phyllis Moen. 2004. "Intimate Academics: Coworking Couples in Two Universities." *Innovative Higher Education* 28: 252–274.

———. 2006. "Advancing a Career Focus on Work and Family: Insights from the Life Course Perspective." Pp. 189–208 in *The Work and Family Handbook: Multi-Disciplinary Perspectives, Methods, and Approaches,* edited by Marcie Pitt-Catsouphes, Ellen Ernst Kossek, and Stephen Sweet. Mahwah, NJ: Erlbaum.

———. 2007. "Integrating Educational Careers in Work and Family: Women's Return to School and Family Life Quality." *Community, Work & Family* 10: 233–252.

Sweet, Stephen, Phyllis Moen, and Peter Meiksins. 2007. "Dual Earners in Double Jeopardy: Preparing for Job Loss in the New Risk Economy." Pp 437–461 in *Workplace Temporalities, vol. 17, New Directions in the Sociology of Work,* edited by Beth Rubin. New York: Elsevier.

Sweet, Stephen, Raymond Swisher, and Phyllis Moen. 2006. "Selecting and Assessing the Family Friendly Community: Adaptive Strategies of Middle Class Dual-Earner Couples." *Family Relations.*

Swidler, Ann. 1986. "Culture in Action: Symbols and Strategies." *American Sociological Review* 51: 273–86.

Swisher, Raymond, Stephen A. Sweet, and Phyllis Moen. 2005. "The Family-Friendly Community and Its Life Course Fit for Dual-Earner Couples." *Family Relations* 54: 596–906.

Takaki, Ronald. 1993. *A Different Mirror: A History of Multicultural America.* Boston: Little, Brown.

Tarnoff, Curt, and Larry Nowels. 2004. "Foreign Aid: An Introductory Overview of U.S. Programs and Policy," vol. 98–916, edited by Congressional Research Service. Washington, DC: Library of Congress.

Taylor, Frederick Winslow. 1964 [1911]. *The Principles of Scientific Management.* New York: Harper.

Taylor, Phil, and Peter Bain. 2005. "India Calling to the Far Away Towns: The Call Centre Labour Process and Globalization." *Work, Employment & Society* 19: 261–282.

Thernstrom, Stephan. 1980. *Poverty and Progress.* Cambridge: Harvard University Press.

Thomas, William, and Dorothy Swaine Thomas. 1928. *The Child in America.* Chicago: University of Chicago Press.

Thomas, William I., and Florian Znaniecki. 1958. *The Polish Peasant in Europe and America.* New York: Dover.

Thompson, E. P. 1963. *The Making of the English Working Class.* New York: Pantheon Books.

———. 1967. "Time, Work-Discipline, and Industrial Capitalism." *Past and Present* 38: 56–97.

Thurow, Lester C. 1992. *Head to Head: The Coming Economic Battle Among Japan, Europe, and America.* New York: William Morrow.

———. 1999. "Jobless Figures Deceptive." *Boston Globe* (April 20), p. C4.

Tiano, Susan. 1994. *Patriarchy on the Line: Labor, Gender, and Ideology in the Mexican Maquila Industry.* Philadelphia: Temple University Press.

Tilly, Chris. 1995. *Half a Job: Good and Bad Part-Time Jobs in a Changing Labor Market.* Philadelphia: Temple University Press.

Tocqueville, Alexis de. 1969 [1836]. *Democracy in America.* New York: Doubleday.

Tracy, James F. 1999. "Whistle While You Work: The Disney Company and the Global Division of Labor." *Journal of Communication Inquiry* 23: 374.

Trattner, Walter. 1999. *From Poor Law to Welfare State.* New York: Free Press.

Tseming, Yang. 2004. "The Effectiveness of the NAFTA Environmental Side Agreement's Citizen Submission Process: A Case Study of the Metales y Derivados Matter" (October 15). Available at SSRN: http://ssrn.com/abstract=552483 or DOI: 10.2139/ssrn.552483

Tucker, James. 1993. "Everyday Forms of Employee Resistance." *Sociological Forum* 8: 25–46.

Tucker, Lee. 2000. *Fingers to the Bone: United States Failure to Protect Child Farmworkers.* Washington, DC: Human Rights Watch.

Tulin, Roger. 1984. *A Machinist's Semi-Automated Life.* San Pedro: Singlejack Books.

Uchitelle, Louis. 2006. *The Disposable American: Layoffs and Their Consequences.* New York: Knopf.

Uggen, Christopher, and Amy Blackstone. 2004. "Sexual Harassment as a Gendered Expression of Power." *American Sociological Review* 69: 64–92.

Ulrich, Laurel. 1982. *Good Wives: Image and Reality in the Lives of Women in Northern New England 1650–1750*. New York: Oxford University Press.

United Nations Statistics Division. 2005. "Demographic and Social Indicators: Statistics and Indicators on Women and Men." Retrieved from http://unstats.un.org/unsd/demographic/products/indwm/wwpub.htm

United Nations. 1989. "Convention on the Rights of the Child." vol. A/RES/44/25, edited by U.N. General Assembly. New York: United Nations.

———. 2005. *Population Challenges and Development Goals*. New York: United Nations.

U.S Census Bureau. 2004. "Educational Attainment in the United States in 2003." Washington, DC: U.S. Department of Commerce.

U.S. Department of Labor. 2007. "Unemployment Compensation: A Federal-State Partnership." Retrieved from http://workforcesecurity.doleta.gov/unemploy/pdf/partnership 2007.pdf

Valian, Virginia. 1998. *Why So Slow? The Advancement of Women*. Cambridge, MA: MIT Press.

Vallas, Steven. 2003a. "The Adventures of Managerial Hegemony: Teamwork, Ideology, and Worker Resistance." *Social Problems* 50: 204–225.

———. 2003b. "Rediscovering the Color Line Within Work Organizations: The 'Knitting of Racial Groups' Revisited." *Work and Occupations* 30: 379–400.

Vallas, Steven, and John Beck. 1996. "The Transformation of Work Revisited: The Limits of Flexibility in American Manufacturing." *Social Problems* 43: 339–361.

Valletta, Robert. 2000. "Declining Job Security." In *On the Job: Is Long-Term Employment a Thing of the Past?* Edited by David Neumark. New York: Russell Sage Foundation.

Varma, Roli. 2006. *Harbingers of Global Change: India's Techno-Immigrants in the United States*. Lanham, MA: Lexington Books.

Vaughan, Diane. 2006. "Air Traffic Control Today: Politics, Labor History, and Cultural Reproduction." *Critical Solidarity, Newsletter of the Labor and Labor Movements Section of the American Sociological Association* 6.

Veblen, Thorstein. 1994 [1899]. *The Theory of the Leisure Class*. New York: Penguin Classics.

Vosler, Nancy R., and Deborah Page-Adams. 1996. "Predictors of Depression among Workers at the Time of a Plant Closing." *Journal of Sociology and Social Welfare* 23: 25–42.

Voydanoff, Patricia. 2007. *Work, Family, and Community: Exploring Interconnections*. Mahwah, NJ: Erlbaum.

Wagar, Terry H. 2001. "Consequences of Work Force Reduction." *Journal of Labor Research* 22: 851–862.

Waldinger, Roger. 2001. "Up From Poverty? 'Race,' Immigration and the Fate of Low-Skilled Workers." Pp. 80–116 in *Strangers at the Gates: New Immigrants in Urban America*, edited by Roger Waldinger. Berkeley: University of California Press.

Waldinger, Roger, and Claudia Der-Martirosian. 2000. "Immigrant Workers and American Labor: Challenge or Disaster?" In *Organizing Immigrants: The Challenge for Unions in Contemporary California,* edited by Ruth Milkman. Ithaca, NY: Cornell University Press.

———. 2001. "The Immigrant Niche: Pervasive, Persistent, Diverse." Pp. 228–271 in *Strangers at the Gates: New Immigrants in Urban America,* edited by Roger Waldinger. Berkeley: University of California Press.

Waldinger, Roger, and Michael Lichter. 2003. *How the Other Half Works: Immigration and the Social Organization of Labor.* Berkeley: University of California Press.

Waldinger, Roger, Nelson Lim, and David Cort. 2007. "Bad Jobs, Good Jobs, No Jobs? The Employment Experience of the Mexican-American Second Generation." *Journal of Ethnic and Migration Studies* 33: 1–35.

Wallerstein, Immanuel. 1979. *The Capitalist World Economy.* Cambridge, UK: Cambridge University Press.

———. 1983. *Historical Capitalism.* London: Verso.

Wanberg, Connie R., L. W. Bunce, and Mark B. Gavin. 1999. "Perceived Fairness of Layoffs Among Individuals Who Have Been Laid Off." *Personnel Psychology* 52(1): 59–84.

Wandner, Stephen, and Andrew Stettner. 2000. "Why Are Many Jobless Workers Not Applying for Benefits?" *Monthly Labor Review* 123: 21–32.

Warren, Elizabeth, and Amelia Warren Tyagi. 2003. *The Two-Income Trap: Why Middle-Class Mothers and Fathers Are Going Broke.* New York: Basic.

Weber, Max. 1998 [1905]. *The Protestant Ethic and the Spirit of Capitalism.* Los Angeles: Roxbury.

Weller, Christian, and Edward Wolff. 2005. *Retirement Income: The Crucial Role of Social Security.* Washington, DC: Economic Policy Institute.

Wells, Miriam. 2000. "Immigration and Unionization in the San Francisco Hotel Industry." In *Organizing Immigrants: The Challenge for Unions in Contemporary California,* edited by Ruth Milkman. Ithaca, NY: Cornell University Press.

Wenneras, Christine, and Agnes Wold. 1997. "Nepotism and Sexism in Peer-Review." *Nature* 387: 341–343.

Westman, Mina, Dalia Etzion, and Esti Danon. 2001. "Job Insecurity and Crossover of Burnout in Married Couples." *Journal of Organizational Behavior* 22: 467–481.

Wharton, Carol. 2002. *Framing a Domain for Work and Family: A Study of Women in Residential Real Estate Sales Work.* Lanham, MA: Lexington Books.

Wharton, Carol S. 1994. "Finding Time for the 'Second Shift': The Impact of Flexible Work Schedules on Women's Double Days." *Gender & Society* 8: 189–205.

Whyte, William H. 1956. *The Organization Man.* New York: Simon & Schuster.

Wiesenfeld, Batia, Joel Brockner, and Christopher L. Martin. 1999. "A Self-Affirmation Analysis of Survivors' Reactions to Unfair Organizational Downsizing." *Journal of Experimental Social Psychology* 35(5): 441–460.

Williams, Christine. 1991. *Gender Differences at Work: Women and Men in Nontraditional Occupations.* Los Angeles: University of California Press.

Williams, Christine, Patti Giuffre, and Kirsten Dellinger. 1999. "Sexuality in the Workplace: Organizational Control, Sexual Harassment, and the Pursuit of Pleasure." *Annual Review of Sociology* 25: 73–93.

Williams, Joan. 2000. *Unbending Gender: Why Family and Work Conflict and What To Do About It.* New York: Oxford University Press.

Wilson, George. 1997. "Pathways to Power: Racial Differences in the Determinants of Job Authority." *Social Problems* 44: 38–54.

Wilson, Kenneth, and Alejandro Portes. 1980. "Immigrant Enclaves: An Analysis of the Labor Market Experiences of Cubans in Miami." *American Journal of Sociology* 86: 295–319.

Wilson, William Julius. 1978. *The Declining Significance of Race: Blacks and Changing American Institutions.* Chicago: University of Chicago Press.

———. 1987. *The Truly Disadvantaged: The Inner City, the Underclass, and Public Policy.* Chicago: University of Chicago Press.

———. 1997. *When Work Disappears: The World of the New Urban Poor.* New York: Knopf.

Winstanley, D., J. Clark, and H. Leeson. 2002. "Approaches to Child Labour in the Supply Chain." *Business Ethics: A European Review* 11: 210–223.

Wolkinson, Benjamin, and Russell Ormiston. 2006. "The Arbitration of Work-Family Conflicts." In *The Handbook of Work and Family: Multi-Disciplinary Perspectives, Methods, and Approaches,* edited by Marcie Pitt-Catsouphes, Ellen Ernst Kossek, and Stephen Sweet. Mahwah, NJ: Erlbaum.

Wood, Ellen Meiksins. 2003. *Empire of Capital.* London: Verso.

Woods, James, and Jay Lucas. 1993. *The Corporate Closet: The Professional Lives of Gay Men in America.* New York: Free Press.

Woods, Ngaire. 2006. *The Globalizers: The IMF, the World Bank, and Their Borrowers.* Ithaca, NY: Cornell University Press.

Wright, Eric Olin. 1985. *Classes.* London: Verso.

Wrigley, Julia. 1999. "Is Racial Oppression Intrinsic to Domestic Work? The Experiences of Children's Caregivers in Contemporary America." Pp. 97–123 in *The Cultural Territories of Race: Black and White Boundaries,* edited by Michele Lamont. Chicago: University of Chicago Press.

Wrong, Dennis. 1961. "The Oversocialized Concept of Man in American Sociology." *American Sociological Review* 26.

Wypijewski, JoAnn. 2006. "Workless Blues." *New Left Review* 42: 141–149.

Yakura, Elaine. 2001. "Billables: The Valorization of Time in Consulting." *American Behavioral Scientist* 44: 1076–1095.

York, Richard, Eugene Rosa, and Thomas Dietz. 2003. "Footprints on the Earth: The Environmental Consequences of Modernity." *American Sociological Review* 68: 279–300.

Young, Alford. 1999. "Navigating Race: Getting Ahead in the Lives of 'Rags to Riches' Young Black Men." Pp. 30–62 in *The Cultural Territories of Race: Black and White Boundaries,* edited by Michèle Lamont. Chicago: University of Chicago Press.

———. 2003. *The Minds of Marginalized Black Men: Making Sense of Mobility, Opportunity, and Future Life Chances.* Princeton, NJ: Princeton University Press.

Zaretsky, Eli. 1986. *Capitalism, the Family, and Personal Life.* New York: Harper & Row.

Zelizer, Viviana. 1994. *Pricing the Priceless Child: The Changing Social Value of Children.* Princeton, NJ: Princeton University Press.

Zeng, Zhen, and Yu Xie. 2004. "Asian-Americans' Earnings Disadvantage Reexamined: The Role and Place of Education." *American Journal of Sociology* 109: 1075–1108.

Zimmer, Lynn. 1986. *Women Guarding Men.* Chicago: University of Chicago Press.

Zuboff, S. 1988. *In the Age of the Smart Machine.* New York: Harper Collins.

Zweigenhaft, Richard, and William Domhoff. 1998. *Diversity in the Power Elite: Have Women and Minorities Reached the Top?* New Haven, CT: Yale University Press.

———. 2003. *Blacks in the White Elite: Will the Progress Continue?* Lanham, MD: Rowman & Littlefield.

Index